# Solar Energy Applications in Houses, Smart Cities and Microgrids

# Solar Energy Applications in Houses, Smart Cities and Microgrids

Special Issue Editor

**Luis Hernández Callejo**

MDPI • Basel • Beijing • Wuhan • Barcelona • Belgrade

**MDPI**

*Special Issue Editor*
Luis Hernández Callejo
Universidad de Valladolid
Spain


*Editorial Office*
MDPI
St. Alban-Anlage 66
4052 Basel, Switzerland


This is a reprint of articles from the Special Issue published online in the open access journal *Applied Sciences* (ISSN 2076-3417) in 2019 (available at: https://www.mdpi.com/journal/applsci/special_issues/solar_energy_applications_houses_smart_cities_microgrids).


For citation purposes, cite each article independently as indicated on the article page online and as indicated below:

LastName, A.A.; LastName, B.B.; LastName, C.C. Article Title. *Journal Name* **Year**, *Article Number*, Page Range.


**ISBN 978-3-03928-068-1 (Pbk)**
**ISBN 978-3-03928-069-8 (PDF)**

# Contents

# About the Special Issue Editor

**Luis Hernández Callejo** has published numerous papers in Q1 and Q2 journals, in first positions in their respective areas. His publications are cited an average of 10 times per year. He has also participated in conferences, seminars, and conferences. He is a reviewer for scientific journals (Energies, Sensors, IEEE Communications Magazine, Energy, etc.). Dr. Luis Hernández has been the coordinator of R & D projects at CEDER-CIEMAT in the field of smart grid/smart metering/distributed generation/microgrid. The projects were focused on the intelligent electrical measurement and the integration of renewable generation sources in electrical networks and their control, as well as the communications protocols by which the industry is supported at the moment. CIEMAT created the Virtual Energy Unit, of which Dr. Luis Hernández was the coordinator, whose objective was to join the efforts of different CIEMAT research groups that belonged to different research units (energy, technology, etc.). He was also a member of the FutuRed Governing Group until 1 September 2015. Dr. Luis Hernández studied at the University of Valladolid, earning a degree in agricultural and energy engineering, which included the following topics: hydraulics and energy, wind energy, solar energy, and technological transfer of renewable sources and electrical networks. He also earned a master's degree in bioenergy and energy sustainability, which covered the following topics: sustainable wind energy I + D + i, geothermal energy hydroelectric and marine sustainable I + D + i, sustainable solar energy I + D + i, and microgrids. He was also a teacher–tutor at the National University of Distance Education (UNED). His degree in information technology touched on the following subjects: data structure and algorithms, programming strategies and data function, programming languages, concurrent programming, declarative programming, programming II, compilers, programming and advanced data structure, programming fundamentals, oriented programming objects, computer engineering III, computer networks, and fundamentals of artificial intelligence; his degree in electrical and/or electronic engineering touched on the following subjects: theory of circuits, electrical and magnetic materials, components and electronic circuits, and fields and waves. In addition, Dr. Luis Hernández is a reviewer of research projects and a reviewer for scientific journals with a high impact factor.

# Preface to "Solar Energy Applications in Houses, Smart Cities and Microgrids"

This book will present the reality of renewable energy sources, specifically those based on solar energy. This reality focuses on the integration of solar energy in cities (smart cities), buildings (smart buildings), microgrids, smart rural grids, and other similar environments. The applicability of solar energy is fundamental for the sustainable development of previous infrastructures. Therefore, the topics of interest in this book are as follows:

1. The integration of photovoltaic systems in cities, buildings, and microgrids;
2. The hybridization of photovoltaic systems with electrical storage;
3. The hybridization of photovoltaic systems with other energy sources;
4. Monitoring tools for systems based on solar energy;
5. Tools to improve efficiency in systems based on solar energy;
6. The integration of thermal systems in cities, buildings, and microgrids;
7. The prediction of solar resources for the estimation of small-scale photovoltaic production;
8. Artificial intelligence applied to the management and operation of solar systems.

**Luis Hernández Callejo**
*Special Issue Editor*

*applied sciences*

MDPI

*Article*

# Simulation of a Solar-Assisted Air-Conditioning System Applied to a Remote School

Jesús Armando Aguilar-Jiménez [1], Nicolás Velázquez [1,*], Ricardo López-Zavala [1,2], Luis A. González-Uribe [2], Ricardo Beltrán [3] and Luis Hernández-Callejo [4,*]

[1]   Center for Renewable Energy Studies, Engineering Institute, Autonomous University of Baja California, 21280 Mexicali, Mexico
[2]   Faculty of Engineering, Autonomous University of Baja California, 21280 Mexicali, Mexico
[3]   Department of Environment and Energy, Advanced Materials Research Center, 31136 Chihuahua, Mexico
[4]   Department of Agricultural Engineering and Forestry, Campus Universitario Duques de Soria, University of Valladolid (UVA), 42003 Soria, Spain
*   Correspondence: nicolas.velazquez@uabc.edu.mx (N.V.); luis.hernandez.callejo@uva.es (L.H.-C.)

Received: 6 June 2019; Accepted: 15 August 2019; Published: 18 August 2019

**Abstract:** In this work, we present an absorption cooling system with 35 kW capacity driven by solar thermal energy, installed in the school of Puertecitos, Mexico, an off-grid community with a high level of social marginalization. The cooling system provides thermal comfort to the school's classrooms through four 8.75-kW cooling coils, while a 110-$m^2$ field of evacuated tube solar collectors delivers the thermal energy needed to activate the cooling machine. The characteristics of the equipment installed in the school were used for simulation and operative analysis of the system under the influence of typical factors of an isolated coastal community, such as the influence of climate, thermal load, and water consumption in the cooling tower, among others. The aim of this simulation study was to determine the best operating conditions prior to system start-up, to establish the requirements for external heating and cooling services, and to quantify the freshwater requirements for the proper functioning of the system. The results show that, with the simulated strategies implemented, with a maximum load operation, the system can maintain thermal comfort in the classrooms for five days of classes. This is feasible as long as weekends are dedicated to raising the water temperature in the thermal storage tank. As the total capacity of the system is distributed in the four cooling coils, it is possible to control the cooling demand in order to extend the operation periods. Utilizing 75% or less of the cooling capacity, the system can operate continuously, taking advantage of stored energy. The cooling tower requires about 750 kg of water per day, which becomes critical given the scarcity of this resource in the community.

**Keywords:** renewable energy; solar cooling; isolated community; absorption chiller; TRNSYS

## 1. Introduction

Renewable energies are an excellent option to provide electrical energy services, cooling, heating, food dehydration, and desalination, among others, in remote places where traditional fuels are not available or are difficult to acquire [1]. Decentralized low-capacity electrical systems installed in consumption centers, also known as microgrids, enable services to be brought to isolated regions where traditional electrification technologies are difficult to access [2] and, based on a combination with renewable energy sources, prevent the continued emission of large quantities of pollutants into the environment.

On the other hand, solar energy can be used for the direct production of electrical energy, as well as for heating and/or cooling of spaces or products [3]. If the air conditioning of spaces is required, it can be done using a thermal or electrical source by means of solar collectors or photovoltaic modules, respectively, as well as some source based on fossil fuels. However, thinking of sustainable development of marginalized and isolated communities, traditional fuels must make way for renewable energies, both for reducing pollutants and saving money. Because of this, the cooling and/or air-conditioning systems driven by clean energies attracted attention for their use in isolated communities, giving comfort conditions to the population and increasing their quality of life [4–6]. Absorption cooling technologies are an attractive option for the climatization of spaces because they can be driven by low-temperature thermal energy [7] without the use of large amounts of electrical energy [8], being able to use solar heating technologies as sources of activation [9]. In addition, they have great technological maturity, economic profitability, and high efficiency [10], as well as a great possibility of energy integration with other technologies to offer other services simultaneously, such as desalination [11].

A large number of studies related to solar-assisted absorption cooling technologies were carried out. Aliane et al. [12] conducted an investigation of experimental results and experiences presented in the state of the art of these types of systems, mentioning that it is necessary to have experience in the installation and to keep the design as simple as possible to ensure efficient operation. Bataineh and Taamneh [13] mentioned that low performance and high cost are the main disadvantages of sorption technologies; however, given the synchrony of solar radiation with cooling needs, these systems are still attractive when coupled with solar energy; thus, research continues to seek a solution to their technological, economic, and environmental problems. Lazzarin and Noro [14] mentioned that cooling systems coupled with photovoltaic modules consume half of the investment cost compared to cooling systems driven by thermal energy; thus, a large reduction in the cost of solar collectors must be achieved for them to be competitive.

Soto and Rivera [15] developed an absorption air-conditioning system cooled by air, which works using an ammonia–lithium nitrate mixture. Their experimental results showed that it is possible to cool the absorption system using only air; thus, there is no need for a cooling tower. They achieved a coefficient of performance (COP) in the range of 0.1 to 0.33 with cooling capacities in the order of 0.8 kW to 3.4 KW. However, the effect on electricity consumption due to the motors of the heat sink fans was not mentioned. Chen et al. [16] carried out experimental tests on an air-cooled solar-assisted absorption cooling system applied to residential buildings. They mentioned that, with this type of system, it is possible to save water, as well as reduce maintenance needs and space due to the absence of cooling tower. In addition, they commented that there are still no low-capacity commercial air-cooled absorption chillers, mainly due to the risk of crystallization of the working mix under environmental conditions. Huang et al. [17] presented the results of a 35-kW capacity solar thermal cooling system that began operation in 2018, which was evaluated during the summer cooling and winter heating periods. The average annual COP was in the range of 0.68 and 0.76, while the solar fraction for heating and cooling averaged 56.6 and 62.5%, respectively, generating electrical savings of 10,158 kWh annually. In another experimental study of thermosolar cooling, Rosiek [18] implemented operational strategies to optimize the exergetic efficiency of the system, such as the best absorption chiller–heat source coupling and equipment control. Lubis et al. [19] tested a 239-kW-capacity chiller activated in hybrid manner with solar energy and gas. By characterizing their system under a wide range of operating conditions, they simulated it under a tropical climate scenario of the Asian region, showing good operating potential. Sokhansefat et al. [20] performed a simulation and experimental validation of a five-ton solar absorption cooling system installed in Tehran, Iran. They analyzed different factors that affect the performance of the system and established the optimal sizes of the equipment to achieve a 28% increase in its performance. Li et al. [21] investigated the performance of a 23-kW solar absorption cooling system driven by a parabolic trough collector field for cooling a 102-m$^2$ meeting room. They attributed the low performance of the system to the fact that the pipes were too long and the heat loss was too high.

One of the main tools for the analysis of thermosolar absorption systems is the TRNSYS software; a large number of simulation studies used it to determine the best operating conditions, sizing, and optimizations [22–27], given the advantages of this software when analyzing the systems under conditions closer to reality. In a work related to the use of these systems in schools, Praene et al. [6] presented modeling and simulation studies, as well as preliminary experimental results, of a solar-driven 30-kW LiBr–H$_2$O system installed at a university in France. They performed a complete analysis of the thermal loads of the classrooms using the TRNSYS software during the periods from 8:00 to 12:00 a.m. and from 1:00 to 5:00 p.m., corresponding to the class schedule. Uçkan and Yousif [28], using the TRNSYS software, analyzed the implementation of a 35-kW LiBr–H$_2$O cooling system driven by solar thermal energy in the arid region of northern Iraq, seeking to reduce the consumption of fossil fuels and promote the use of renewables. In another similar study, Dakheel et al. [29] simulated the use of renewable energy-saving technologies in the United Arab Emirates, including solar–thermal absorption cooling systems. Reductions of 19.35% in annual cooling energy use, and 7.2% reduction in total annual energy use were achieved. Abrudan et al. [30] studied the implementation of absorption cooling equipment under different climatic conditions. They proposed new correlations between the solar hot water temperature and the cooling water temperature in order to avoid both crystallization and the reduction of the degassing zone below 6%. They concluded that, depending on the present climatic conditions, it is necessary to implement different operative strategies looking for greater efficiency of the system. Analyzing options to improve the efficiency of this type of system, Bellos and Tzivanidis [31,32] studied the incorporation of nanofluids into solar heating technologies. They found that the use of nanoparticles increased the exergetic efficiency of the collector by about 4% and the cooling output by about 0.84% on a daily basis. Mendecka et al. [33] analyzed two thermal storage options for adsorption cooling systems using solar energy. Their results showed that energy storage using phase change materials was slightly more efficient than storage using water. Another option for the activation of absorption systems is hybrid photovoltaic/thermal (PV/T) technology. Alobaid et al. [34], in their review of the state of the art of these systems, identified that up to 50% primary energy can be saved by using hybrid collectors in absorption machines compared to mechanical vapor compression equipment for cooling. The hybridization of mechanical vapor compression technologies was even proposed, working in combination with absorption technologies to improve the system efficiency [35,36].

It should be noted that very few studies mentioned the benefits of this type of solar-assisted cooling system in applications with difficult access to electricity, such as isolated communities. By using an energy resource that can be exploited in an excellent way in most of the world's territory, it makes its implementation attractive in this type of scenario where there is no electricity supply. This type of application presents factors that affect the operation of the system, different from those that would be expected in absorption systems installed in towns or cities with access to continuous fresh water and electricity services. However, as mentioned in the previous works, a detailed analysis of the engineering of these systems must be carried out so that they become both technically and economically attractive. This work presents the simulation study of an absorption air-conditioning system driven by a solar thermal collector field, installed in the school of Puertecitos, an off-grid community with a high level of social marginalization. The system was simulated using the TRNSYS software in order to determine its performance under the typical operating conditions of a school in Mexico, as well as to determine the effect of different distinctive factors of a remote community, such as the scarcity of clean water and electricity. The aim of this simulation study was to determine the best operating conditions prior to system start-up, to establish the requirements for external heating and cooling services, and to quantify the freshwater requirements for the proper functioning of the system. This study allows establishing the operation and maintenance strategies, as well as the operating limits, of the cooling system under a real scenario that was not studied in the literature.

## 2. System Description

The thermosolar absorption air-conditioning system installed in the primary school of the community of Puertecitos, Baja California, Mexico (30°21′19.7″ north (N), 114°38′26.3″ west (W)) is composed of a field of solar thermal collectors, containing evacuated tubes with a parabolic reflector in the lower part with a total aperture area of 110 m$^2$; these solar heating technologies are responsible for maintaining optimal temperature conditions using a water storage tank of 12 m$^3$ capacity, necessary for proper operation of the cooling system. Both the collector field and the thermal energy storage tank (TEST) compose the hot water circuit.

An LiBr–H$_2$O absorption cooling system of 35 kW capacity is thermally driven by the hot circuit fluid. Due to the absorption process, heat is removed from the water that is used as a cooling medium in the chilled water circuit. For abrupt temperature changes not to affect the operation of the chiller, a 1-m$^3$ buffer tank is used in the chilled water circuit. The chilled water passes through four cooling coils with a capacity of 8.75 kW each, one in each classroom of the school, removing the thermal load when required. The heat supplied to the chiller, both by the hot and chilled water circuit, is discharged through the cooling water circuit and released into the environment with an evaporative cooling tower.

Given the problems of water availability in the region and, being a community next to the sea, the cooling water circuit has two variants: normal water mode and seawater mode. In the first mode of operation, freshwater is used as a direct cooling medium for the chiller; it is fed into the absorption machine to remove the heat and transfer it to the environment with the cooling tower. When seawater is used, the cooling of the system is carried out indirectly. The cooling water circuit is divided into two; the tower is responsible for cooling the seawater and recirculates it in an independent circuit, while, in the other circuit, freshwater is used to remove heat from the chiller and, by means of a titanium coil submerged in the seawater reservoir of the cooling tower, the removed heat is transferred. This is done in order to not introduce the incrusting and corrosive compounds that the seawater carries with it into the internal heat exchangers of the absorption machine, as well as to not depend on replenishment of freshwater due to evaporation by cooling the system. There is also a diesel auxiliary heater in the hot water circuit so that, if the temperature conditions in the TEST are not met, the system can be operated. The previous equipment installed in the Puertecitos school is shown in Figures 1 and 2.

**Figure 1.** Evacuated tube solar collector field with a bottom parabolic reflector.

The schematic diagram of the components and connections is shown in Figure 3. The solar collector field is responsible for maintaining the temperature of the TEST in the range of 75–96 °C, enough for proper chiller operation. Regardless of whether the cooling machine is operating or not (for example, on weekends, the chiller does not operate because there are no classes), the solar field pump continues to provide fluid for heating until the tank is completely at the set point temperature, in this case 96 °C, always seeking to have the system in operating conditions. When there is a need for classroom air conditioning, the chiller turns on and the TEST pump supplies hot water from the top. Since the temperature of the generator of the absorption machine is at an optimal level for operation, the pump of the chilled water circuit is turned on to bring it to the cooling coils of the classrooms, at a temperature between 7 and 10 °C, removing the heat from the spaces to be air-conditioned. This current is taken to the chiller to be cooled again. It is worth mentioning that the configuration of the chilled water circuit allows controlling the number of cooling coils used for air conditioning, allowing a saving of thermal energy when air conditioning is not required in all classrooms.

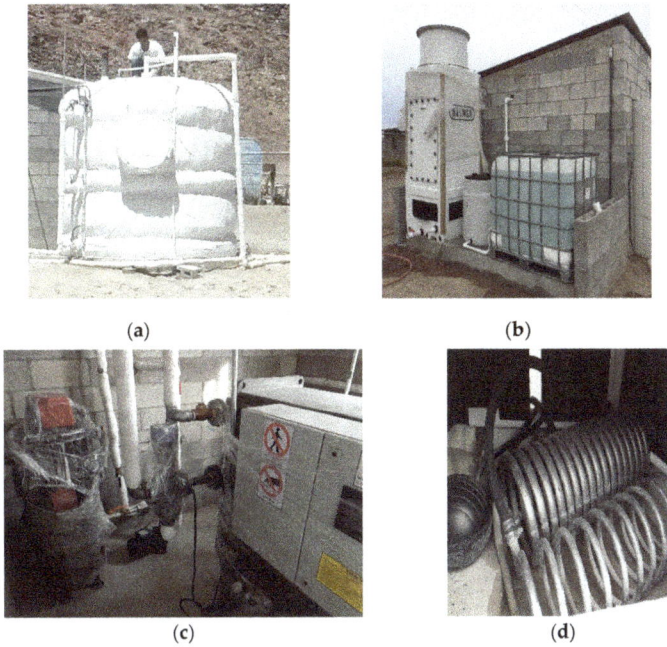

(a)

(b)

(c)

(d)

**Figure 2.** Components of the thermosolar absorption cooling system at Puertecitos school, where we can appreciate the (**a**) thermal energy storage tank, (**b**) cooling tower with freshwater and seawater storage tanks, (**c**) auxiliary heater and chiller inside the machine room, and (**d**) titanium coil inside the cooling tower.

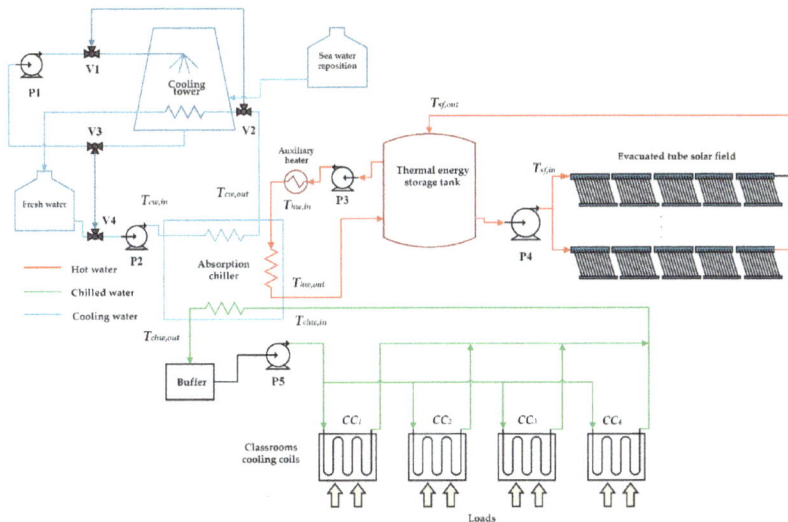

**Figure 3.** Schematic diagram of the equipment and circuits of the solar thermal cooling system.

Depending on the mode of operation in the cooling tower mentioned above, the cooling water circuit is modified with manual valves for the use of seawater and/or freshwater only. When seawater is required as cooling fluid in the tower, the seawater reposition tank fills the cooling tower tank while the P1 pump recirculates the fluid in a closed circuit between the V1 and V3 valves, cooling the seawater by evaporation. During this process, the P2 pump takes water from the freshwater tank and supplies it to the chiller to remove the heat generated. Then, it transfers it to the seawater by means of a coil heat exchanger submerged in the cooling tower reservoir (Figure 2d) and returns it to the freshwater tank, all in the closed circuit between V2 and V4. This configuration avoids the problem of freshwater evaporation in the cooling tower by using seawater, an easier resource to acquire in the community. On the other hand, when freshwater is used as a direct cooling medium, the P2 pump takes water from the cooling tower reservoir through the V3–V4 circuit, it removes the heat from the chiller, and, through the V2–V1 circuit, it cools the freshwater through evaporation in the tower. The freshwater tank is positioned in a way that when the water level in the tower tank decreases, the water is replaced by gravity and controlled by a float. In the latter configuration, the P1 pump does not work.

## 3. Methodology

The TRNSYS software was used to carry out the operational simulation of the solar thermal cooling system. This program allows analyzing the system in a semi-dynamic way, having the advantage of working with typical meteorological databases (TMY) in the region, studying the system under more realistic operating conditions. By using modules programmed with validated mathematical models of various equipment, such as pumps, motors, heat exchangers, storage tanks, solar collectors, power cycles, and refrigeration, among many others, TRNSYS becomes an ideal software for the simulation of this type of process. The validated mathematical models used in the simulator modules can be reviewed in Reference [37]. Table 1 shows the characteristics of the cooling system equipment installed in the Puertecitos school, which were used as input parameters in the developed simulator, and Table 2 presents the description of the components within TRNSYS.

**Table 1.** System characteristics.

| Solar Collector | |
|---|---|
| Brand/model | Suntask/SHC24 |
| Collector type | Evacuated tube with parabolic reflector |
| Number of tubes | 24 |
| Aperture area | $4.41 \text{ m}^2$ |
| Optical efficiency ($a_0$) | 0.668 |
| First order efficiency coefficient ($a_1$) | $1.496 \text{ W/m}^2 \text{ K}$ |
| Second order efficiency coefficient ($a_2$) | $0.005 \text{ W/m}^2 \text{ K}^2$ |
| Fluid | Water |
| Mass flow | $0.02 \text{ kg/sec m}^2$ |
| Number in series | 5 |
| Number of loops | 5 |
| **Thermal Energy Storage Tank** | |
| Volume | $12 \text{ m}^3$ |
| Height | $2.5 \text{ m}$ |
| Material | Fiberglass |
| Insulation thickness | $0.025 \text{ m}$ |
| Loss coefficient | $1.4 \text{ W/m}^2 \text{ K}$ |
| Fluid | Water |

**Table 1.** *Cont.*

| Absorption Chiller | |
|---|---|
| Model | Lucy New Energy/RXZ-35 |
| Refrigerant | Water–lithium bromide |
| Cooling capacity | 35 kW |
| COP | 0.7 |
| Hot water nominal temperature | 90 °C |
| Hot water nominal flow rate | 8.3 m³/h |
| Chilled water nominal inlet temperature | 15 °C |
| Chilled water nominal flow rate | 6 m³/h |
| Cooling water nominal temperature | 30 °C |
| Cooling water nominal flow rate | 15 m³/h |
| Power consumption | 0.3 kW |

**Table 2.** TRNSYS type for the components simulated.

| Component | TRNSYS Type | Description |
|---|---|---|
| Temperature control | 2 | On/off control of feed pumps of the different circuits. |
| Pumps | 3 | Pumps for mass flow feeding of circuits. |
| Time-dependent forcing function | 14 | Cooling system on/off time control. |
| Weather data processor | 15 | Climatological database. |
| Cooling tower | 51 | Evaporative cooling tower with constant volumetric flow. |
| Solar thermal collector | 71 | Evacuated tube solar thermal collector. |
| Absorption chiller | 107 | LiBr/$H_2O$ hot water-driven absorption chiller. |
| Thermal energy storage tank | 534 | Stratified thermal storage tank with variable nodes, ports, and insulation. |
| Buffer tank | 534 | Thermal storage tank with variable nodes, ports, and insulation. |
| Mass flow diverter | 647 | Mass flow diverter with variable outlets. |
| Mass flow mixer | 649 | Mass flow mixer with variable inlets. |
| Mass flow heat exchanger | 682 | Energy and mass balance heat exchanger. |
| Heat load | 686 | Heat load profile dependent on the day, weekday, month, or season. |

In order to carry out the analysis of the system with the aforementioned software, the following considerations were taken into account:

- Steady state;
- 15-min simulation intervals;
- Maximum thermal load of 8.5 kW per classroom at 12:00 p.m.;
- Pressure drops and heat losses in the equipment were not considered, only the heat transferred to the environment by the TEST losses was considered;
- On/off control by temperature difference in the solar field ($T_{sf,out} - T_{sf,in} > 0$);
- Only fresh water was used in the cooling water circuit;
- System operation with solar energy only.

The cooling system operates under an established school schedule of 8:00 a.m. to 3:00 p.m. Monday through Friday, which are typical class hours in Mexico. The vacation periods correspond to the months of July and part of August, the same days in which the highest ambient temperatures are present; thus, a constant operation is expected during May, June, mid-August, and September, where the penultimate month is the critical one due to the extreme temperatures present. Figure 4 shows the ambient temperature, relative humidity, and global radiation conditions based on a TMY file of the study region for the simulation period, corresponding to warm and sunny days in August. Being a coastal community, the relative humidity plays a very important role in the proper operation of the cooling water circuit and, therefore, the system in general.

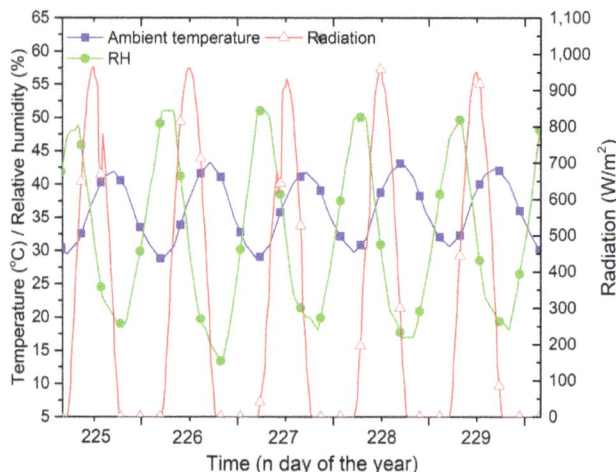

**Figure 4.** Ambient temperature, relative humidity, and global radiation in the simulation lapse for Puertecitos, México.

## 4. Results and Discussion

### 4.1. Operation at Nominal Capacity

Seeking adequate temperature conditions inside the classrooms, the chiller is turned on every day that cooling is needed from 7:00 a.m. to 3:00 p.m., i.e., one hour before the beginning of classes. This results in a constant eight-hour operation from Monday to Friday, allowing robust system programming and control without much user intervention.

Figure 5 shows the variation of the thermal load present in the classrooms during a typical operating day and the temperature of the chilled water coming from the absorption chiller. The thermal load starts at 6:00 a.m.; however, the chiller starts cooling the water at 7:00 a.m. As the hours pass, the load increases until it reaches 35 kW, with the maximum demand established at 12:00 p.m. corresponding to 8.75 kW per cooling coil. At 6:00 p.m., the thermal load is zero, but the cooling system is stopped 3 h beforehand as classes are finished and the facilities are empty, avoiding the consumption of thermal and electrical energy. The chilled water temperature remains constant at 7 °C during chiller operation due to the control of the heat supplied to the system by means of the hot water circuit. If the cooling load increases, the heat is taken out of the TEST to raise the capacity of the chiller and keep the operating conditions as stable as possible. Otherwise, if the load increases but the heat supplied to the system remains constant, it results in an increase in the temperature of the chilled water and, therefore, of the classrooms. In addition, the heat transfer rate to the chiller cooling water must be sufficient to maintain stable operating conditions, so that, as the cooling capacity increases, more heat is removed from the system by the cooling water circuit. It should be noted that the temperatures of the hot water and cooling water circuit do not remain constant, as they depend directly on the environmental conditions present during the operation of the system; however, they are always within the minimum ranges, as shown in Figure 6.

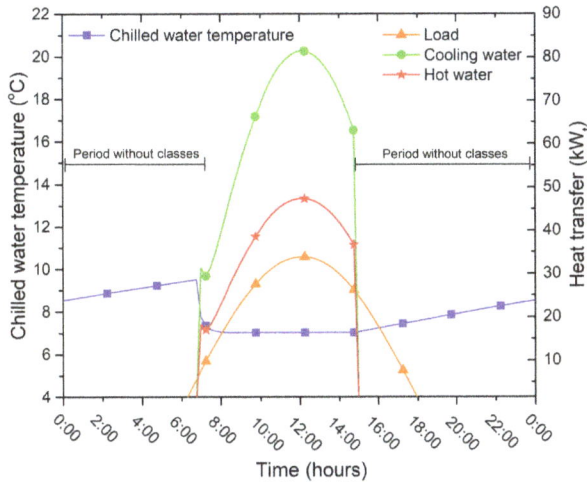

**Figure 5.** Chilled water temperature and heat transfer of the circuits.

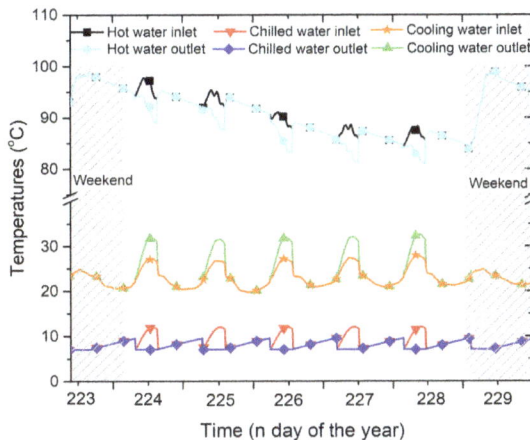

**Figure 6.** Variation in system circuit temperatures.

A typical week of operation is shown in Figure 6, where the temperature variations of the different circuits of the cooling system at its inlet and outlet can be appreciated. At the beginning of the week, the temperature of the TEST is in optimal conditions to start the system, around 97 °C for all its levels, as the field of solar collectors is dedicated to its increase in temperature on Saturday and Sunday. During the days of classes, the TEST tends to decrease its temperature significantly because the collection and storage of daily solar energy is not enough to return it to initial conditions. Since the operation of the chiller stops at 3:00 p.m., while there is still a solar resource, the TEST recovers a bit of its temperature level by keeping the solar collector pump on; however, this is not enough to counteract the loss of heat during the day. With the passing of operating days, the decrease in temperature becomes greater; thus, when the last day of classes arrives, the system operates in limited conditions of activation temperature (around 75–80 °C), while being able to satisfy the cooling demand of the classrooms. However, it is necessary for weekends to be exclusively dedicated to heating the TEST and bringing it to its initial conditions so that, at the beginning of the following week, the complete system is in optimal operating conditions and meets the cooling requirements during the five days of classes.

### 4.2. Partial Load Operation

The school where the thermosolar cooling system is installed has fluctuations in student attendance, mainly due to migration and immigration phenomena of its population, depending on weather or working conditions. For this reason, depending on the number of students in the school, it may not be necessary to air-condition the four classrooms, but only some of them.

Figure 7 shows the behavior of the mean temperature of the TEST with respect to the variation of the maximum thermal load of the classrooms, using the profile presented in Figure 5 at 25, 50, 75, and 100% of the chiller capacity. Working the system with a maximum load of 35 kW ensures operation during the five days of classes as long as weekends are used to heat the TEST, as mentioned in the previous section. However, with a maximum load of 25.5 kW, corresponding to 75% of the capacity of the cooling system, it is possible to maintain constant operation for more than five days without having to spend the entire weekend heating it. The decrease in the average temperature of the TEST is recovered in the period with the availability of the solar resources after typical operation of 7:00 a.m. to 3:00 p.m. On the other hand, when operating at 50% of the maximum capacity or less, the temperature conditions of the system are kept stable, being able to operate normally for the whole week in the aforementioned period.

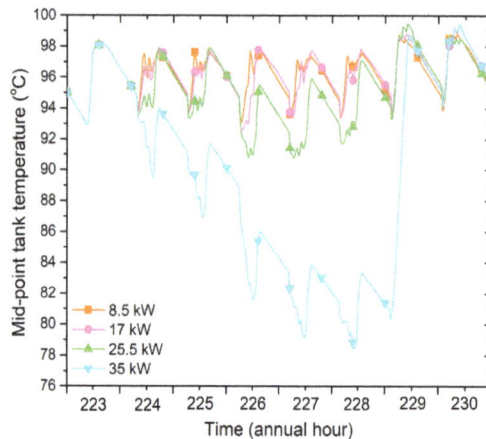

**Figure 7.** Thermal energy storage tank (TEST) temperature when varying the thermal load.

### 4.3. Evaporation of Cooling Water

The availability of water and electricity in this region is a very severe problem. There is no water distribution network and the easiest way to acquire it is in a well located 32 km away from the community; thus, it becomes a natural resource much appreciated and cared for. On the other hand, electricity is supplied by a microgrid based on renewable energy, but the school does not have the economic resources to pay for it. However, the heat entering the cooling system needs to be removed; thus, the use of a cooling tower is essential. There are two options: cooling with water or air. The first consumes a considerable amount of water with low electricity consumption, while the second does not consume water, but a large amount of electricity. Water cooling was selected by installing equipment that allows both seawater and freshwater as a working fluid. Thus, if freshwater is not available, seawater can be used given the ease of acquisition.

Figure 8 shows the weekly water consumption in the cooling tower. Depending on the thermal load of the chiller, the water evaporation rate can reach 140 kg/h. During a typical eight-hour operating day, about 750 kg of water would be needed per day, giving a total of 4000 kg (~4 m$^3$) per normal week of classes. This is critical considering the scarcity of the resource in the region and the work involved in maintaining tank levels in these conditions, as well as maintenance related to using seawater in the cooling tower due to salt concentration. With these results, it is possible to program the filling of the tanks and the cleaning of the cooling tower tank.

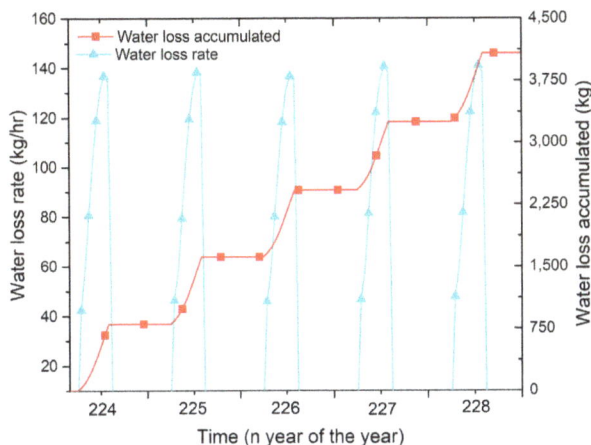

**Figure 8.** Rate of water loss due to evaporation in the cooling tower and its weekly accumulation.

## 5. Conclusions

An operational analysis of a 35-kW absorption cooling system driven by solar thermal energy was carried out and installed in a primary school in the isolated community of Puertecitos, Baja California, Mexico. The system was simulated using TRNSYS software in order to determine its performance under the typical operating conditions of a school in Mexico, as well as to determine the effect of different distinctive factors of a remote community, such as the scarcity of freshwater and electricity. The simulation results show that, under full load operation, the cooling system can satisfy the air-conditioning requirements of the four classrooms during the five days of the week, from 7:00 a.m. to 3:00 p.m. The weekends are used exclusively for the storage of thermal energy and elevation of the temperature of the TEST, using the solar collector field. The operation of the system under different thermal load profiles was also studied, and it was concluded that, under a thermal load of 75% of the chiller capacity or less, the system can operate the entire week without the need to dedicate weekends solely to thermal energy storage. With the configuration in the chilled water circuit, the individual operation of each cooling coil is possible; thus, it is possible to control the thermal load of the classrooms to extend the operation of the chiller system.

On the other hand, the consumption of freshwater in the cooling tower is a critical factor due to the scarcity of this resource in the region. At its maximum capacity, the system consumes about 4 m$^3$ of water to remove heat from the equipment per normal week of operation. Therefore, the system is designed with two independent cooling water circuits, where seawater can be used in the cooling tower as it is easier to acquire in the project community. This study allowed establishing the operation and maintenance strategies, as well as determining the best operating conditions prior to system start-up, establishing the requirements for external heating and cooling services, and quantifying the freshwater requirements for the proper functioning of the system under a real scenario not studied in the literature.

**Author Contributions:** Investigation, J.A.A.-J.; methodology, R.L.-Z., L.H.-C., and L.A.G.-U.; software, J.A.A.-J. and N.V.; formal analysis, J.A.A.-J., R.B., L.H.-C., and N.V.; writing—original draft preparation, J.A.A.-J., L.A.G.-U., and R.L.-Z.; writing—review and editing, N.V., R.B., and L.H.-C.; supervision, N.V. and L.H.-C.; funding acquisition, N.V. and L.H.-C.

**Funding:** This research was funded by CONACYT-SENER-SUSTENTABILIDAD ENERGÉTICA through project P09 of CEMIE-Solar.

**Acknowledgments:** The authors acknowledge CONACYT for the support received through a graduate scholarship for Jesús Armando Aguilar-Jiménez. The authors also acknowledge the CYTED Thematic Network "CIUDADES INTELIGENTES TOTALMENTE INTEGRALES, EFICIENTES Y SOSTENIBLES (CITIES)" n° 518RT0558.

**Conflicts of Interest:** The authors declare no conflicts of interest.

## References

1. Ehsan, A.; Yang, Q. Scenario-based investment planning of isolated multi-energy microgrids considering electricity, heating and cooling demand. *Appl. Energy* **2019**, *235*, 1277–1288. [CrossRef]
2. Aguilar-Jiménez, J.A.; Velázquez, N.; Acuña, A.; Cota, R.; González, E.; González, L.; López, R.; Islas, S. Techno-economic analysis of a hybrid PV-CSP system with thermal energy storage applied to isolated microgrids. *Sol. Energy* **2018**, *174*, 55–65. [CrossRef]
3. Mittelman, G.; Kribus, A.; Dayan, A. Solar cooling with concentrating photovoltaic/thermal (CPVT) systems. *Energy Convers. Manag.* **2007**, *48*, 2481–2490. [CrossRef]
4. Tsoutsos, T.; Aloumpi, E.; Gkouskos, Z.; Karagiorgas, M. Design of a solar absorption cooling system in a Greek hospital. *Energy Build.* **2010**, *42*, 265–272. [CrossRef]
5. Marc, O.; Lucas, F.; Sinama, F.; Monceyron, E. Experimental investigation of a solar cooling absorption system operating without any backup system under tropical climate. *Energy Build.* **2010**, *42*, 774–782. [CrossRef]
6. Praene, J.P.; Marc, O.; Lucas, F.; Miranville, F. Simulation and experimental investigation of solar absorption cooling system in Reunion Island. *Appl. Energy* **2011**, *88*, 831–839. [CrossRef]
7. Pranesh, V.; Velraj, R.; Christopher, S.; Kumaresan, V. A 50 year review of basic and applied research in compound parabolic concentrating solar thermal collector for domestic and industrial applications. *Sol. Energy* **2019**, *187*, 293–340. [CrossRef]
8. Xu, Z.Y.; Wang, R.Z. Comparison of absorption refrigeration cycles for efficient air-cooled solar cooling. *Sol. Energy* **2018**, *172*, 14–23. [CrossRef]
9. Aguilar-Jiménez, J.A.; Velázquez, N.; Acuña, A.; López-Zavala, R.; González-Uribe, L.A. Effect of orientation of a CPC with concentric tube on efficiency. *Appl. Therm. Eng.* **2018**, *130*, 221–229. [CrossRef]
10. Shirazi, A.; Taylor, R.A.; Morrison, G.L.; White, S.D. Solar-powered absorption chillers: A comprehensive and critical review. *Energy Convers. Manag.* **2018**, *171*, 59–81. [CrossRef]
11. López-Zavala, R.; Velázquez-Limón, N.; González-Uribe, L.A.; Aguilar-Jiménez, J.A.; Alvarez-Mancilla, J.; Acuña, A.; Islas, S. A novel LiBr/H2O absorption cooling and desalination system with three pressure levels. *Int. J. Refrig.* **2019**, *99*, 469–478. [CrossRef]
12. Aliane, A.; Abboudi, S.; Seladji, C.; Guendouz, B. An illustrated review on solar absorption cooling experimental studies. *Renew. Sustain. Energy Rev.* **2016**, *65*, 443–458. [CrossRef]
13. Bataineh, K.; Taamneh, Y. Review and recent improvements of solar sorption cooling systems. *Energy Build.* **2016**, *128*, 22–37. [CrossRef]
14. Lazzarin, R.M.; Noro, M. Past, present, future of solar cooling: Technical and economical considerations. *Sol. Energy* **2018**, *172*, 2–13. [CrossRef]
15. Soto, P.; Rivera, W. Experimental assessment of an air-cooled absorption cooling system. *Appl. Therm. Eng.* **2019**, *155*, 147–156. [CrossRef]
16. Chen, J.F.; Dai, Y.J.; Wang, R.Z. Experimental and analytical study on an air-cooled single effect LiBr-H2O absorption chiller driven by evacuated glass tube solar collector for cooling application in residential buildings. *Sol. Energy* **2017**, *151*, 110–118. [CrossRef]
17. Huang, L.; Zheng, R.; Piontek, U. Installation and Operation of a Solar Cooling and Heating System Incorporated with Air-Source. *Energies* **2019**, *12*, 996. [CrossRef]
18. Rosiek, S. Exergy analysis of a solar-assisted air-conditioning system: Case study in southern Spain. *Appl. Therm. Eng.* **2019**, *148*, 806–816. [CrossRef]

19. Lubis, A.; Jeong, J.; Saito, K.; Giannetti, N.; Yabase, H.; Idrus Alhamid, M. Nasruddin Solar-assisted single-double-effect absorption chiller for use in Asian tropical climates. *Renew. Energy* **2016**, *99*, 825–835. [CrossRef]

20. Sokhansefat, T.; Mohammadi, D.; Kasaeian, A.; Mahmoudi, A.R. Simulation and parametric study of a 5-ton solar absorption cooling system in Tehran. *Energy Convers. Manag.* **2017**, *148*, 339–351. [CrossRef]

21. Li, M.; Xu, C.; Hassanien, R.H.E.; Xu, Y.; Zhuang, B. Experimental investigation on the performance of a solar powered lithium bromide–water absorption cooling system. *Int. J. Refrig.* **2016**, *71*, 46–59. [CrossRef]

22. Calise, F.; Dentice d'Accadia, M.; Palombo, A. Transient analysis and energy optimization of solar heating and cooling systems in various configurations. *Sol. Energy* **2010**, *84*, 432–449. [CrossRef]

23. Cascetta, F.; Di Lorenzo, R.; Nardini, S.; Cirillo, L. A Trnsys Simulation of a Solar-Driven Air Refrigerating System for a Low-Temperature Room of an Agro-Industry site in the Southern part of Italy. *Energy Procedia* **2017**, *126*, 329–336. [CrossRef]

24. Shirazi, A.; Pintaldi, S.; White, S.D.; Morrison, G.L.; Rosengarten, G.; Taylor, R.A. Solar-assisted absorption air-conditioning systems in buildings: Control strategies and operational modes. *Appl. Therm. Eng.* **2016**, *92*, 246–260. [CrossRef]

25. Asim, M.; Dewsbury, J.; Kanan, S. TRNSYS Simulation of a Solar Cooling System for the Hot Climate of Pakistan. *Energy Procedia* **2016**, *91*, 702–706. [CrossRef]

26. Khan, M.S.A.; Badar, A.W.; Talha, T.; Khan, M.W.; Butt, F.S. Configuration based modeling and performance analysis of single effect solar absorption cooling system in TRNSYS. *Energy Convers. Manag.* **2018**, *157*, 351–363. [CrossRef]

27. Xu, Y.Z.; Wang, Z.R. Simulation of solar cooling system based on variable effect LiBr-water absorption chiller. *Renew. Energy* **2017**, *113*, 907–914. [CrossRef]

28. Uçkan, I.; Yousif, A.A. Environmental Effects Simulation of a solar absorption cooling system in Dohuk city of the Northern Iraq. *Energy Sources Part A Recover. Util. Environ. Eff.* **2019**, *0*, 1–17. [CrossRef]

29. Al Dakheel, J.; Aoul, K.T.; Hassan, A. Enhancing green building rating of a school under the hot climate of UAE; Renewable energy application and system integration. *Energies* **2018**, *11*, 2465. [CrossRef]

30. Abrudan, A.C.; Pop, O.G.; Serban, A.; Balan, M.C.; Abrudan, A.C.; Pop, O.G.; Serban, A.; Balan, M.C. New Perspective on Performances and Limits of Solar Fresh Air Cooling in Different Climatic Conditions. *Energies* **2019**, *12*, 2113. [CrossRef]

31. Bellos, E.; Tzivanidis, C. Parametric analysis and optimization of an Organic Rankine Cycle with nanofluid based solar parabolic trough collectors. *Renew. Energy* **2017**, *114*, 1376–1393. [CrossRef]

32. Bellos, E.; Tzivanidis, C. Alternative designs of parabolic trough solar collectors. *Prog. Energy Combust. Sci.* **2019**, *71*, 81–117. [CrossRef]

33. Mendecka, B.; Cozzolino, R.; Leveni, M.; Bella, G. Energetic and exergetic performance evaluation of a solar cooling and heating system assisted with thermal storage. *Energy* **2019**, *176*, 816–829. [CrossRef]

34. Alobaid, M.; Hughes, B.; Kaiser, J.; Connor, D.O.; Heyes, A. A review of solar driven absorption cooling with photovoltaic thermal systems. *Renew. Sustain. Energy Rev.* **2017**, *76*, 728–742. [CrossRef]

35. Li, Z.; Yu, J.; Chen, E.; Jing, Y. Off-Design Modeling and Simulation of Solar Absorption-Subcooled Compression Hybrid Cooling System. *Appl. Sci.* **2018**, *8*, 2612. [CrossRef]

36. Jing, Y.; Li, Z.; Liu, L.; Lu, S. Exergoeconomic assessment of solar absorption and absorption-compression hybrid refrigeration in building cooling. *Entropy* **2018**, *20*, 130. [CrossRef]

37. Klein, S.A.; Beckman, W.A.; Mitchell, J.W.; Duffie, J.A.; Duffie, N.A.; Freeman, T.L.; Mitchell, J.C.; Braun, J.E.; Evans, B.L.; Kummer, J.P.; et al. *TRNSYS 17: A Transient System Simulation Program: Mathematical Reference*; University of Wisconsin: Madison, WI, USA, 2009.

*applied*
*sciences*

MDPI

*Article*

# Thermal Evaluation of Graphene Nanoplatelets Nanofluid in a Fast-Responding HP with the Potential Use in Solar Systems in Smart Cities

**M. M. Sarafraz [1], Iskander Tlili [2], Zhe Tian [3], Mohsen Bakouri [4], Mohammad Reza Safaei [5,6,* ] and Marjan Goodarzi [7]**

[1] School of Mechanical Engineering, the University of Adelaide, South Australia 5005, Australia; mohammadmohsen.sarafraz@adelaide.edu.au
[2] Department of Mechanical and Industrial Engineering, College of Engineering, Majmaah University, Al-Majmaah 11952, Saudi Arabia; l.tlili@mu.edu.sa
[3] School of Engineering, Ocean University of China, Qingdao 266100, China; zhetian_ouc@163.com
[4] Department of Medical Equipment Technology, College of Applied Medical Sciences, Majmaah University, Al-Majmaah 11952, Saudi Arabia; m.bakouri@mu.edu.sa
[5] Division of Computational Physics, Institute for Computational Science, Ton Duc Tang University, Ho Chi Minh City 758307, Vietnam
[6] Faculty of Electrical and Electronics Engineering, Ton Duc Thang University, Ho Chi Minh City 758307, Vietnam
[7] Department of Mechanical Engineering, Lamar University, Beaumont, TX 77705, USA; mgoodarzi@lamar.edu
* Correspondence: cfd_safaei@tdtu.edu.vn; Tel.: +1-502-657-9981

Received: 25 April 2019; Accepted: 16 May 2019; Published: 22 May 2019

**Abstract:** An experimental study was undertaken to assess the heat-transfer coefficient (HTC) of graphene nanoplatelets-pentane nanofluid inside a gravity-assisted heat pipe (HP). Influence of various parameters comprising heat flux, mass fraction of the nanoparticles, installation angle and filling ratio (FR) of the working fluid on the HTC of the HP was investigated. Results showed that the HTC of the HP was strongly improved due to the presence of the graphene nanoplatelets. Also, by enhancing the heat flux, the HTC of the HP was improved. Two trade-off behaviors were identified. The first trade-off belonged to the available space in the evaporator and the heat-transfer coefficient of the system. Another trade-off was identified between the installation angle and the residence time of the working fluid inside the condenser unit. The installation angle and the FR of the HP were identified in which the HTC of the HP was the highest. The value of installation angle and filling ratio were 65° and 0.55, respectively. Likewise, the highest HTC was obtained at the largest mass fraction of the graphene nanoplatelets which was at wt. % = 0.3. The improvement in the HTC of the HP was ascribed to the Brownian motion and thermophoresis effects of the graphene nanoplatelets.

**Keywords:** graphene; n-pentane; thermosyphon; Thermal performance; tilt angle; filling ratio

## 1. Introduction

With the continuous advancement and development in the structure of the cities, demand for new technologies and, as a result, energy has increased [1]. With limitation in fossil fuel and environmental pollution due to the combustion of carbonaceous fuels, special attention has been paid to renewable energies as a reliable source of energy [2,3]. Solar energy is one of the potential renewable energy resources, which is available to a vast region of the world. However, the intermittent behavior of solar energy during the on-sun and off-sun periods dictates that the solar receivers and collectors must have plausible thermal efficiency to maximize the energy absorption from the sun [4–6]. A solar thermal collector is a two-phase apparatus embedded with heat pipes (HPs) to absorb the solar thermal energy and transport it to the top header using the HPs [7]. A HP is a heat-exchanging medium, which employs a working fluid with plausible thermal properties with low boiling point which can easily be evaporated inside the HP [8–11]. A HP composed of an "evaporator section", in which the working fluid is evaporated using boiling mechanism and flows through an "adiabatic section" to reach a "condenser section". In the condenser section, the vapor phase loses its thermal energy and is condensed to the liquid phase. The formed liquid returns to the evaporator using a gravity-assisted falling film. The heat transfer in HPs does not need any external energy, thereby nominating HPs as one of the efficient technologies for heat transfer within a confined space. The current HPs employ alcohols and acetone as the working fluid; however, these working fluids have poor thermal conductivity, which strongly affects the thermal performance of the HPs. Since Aragon National Laboratory introduced the "nanofluids", a new direction of research was developed in thermal engineering science and in different research areas including single and two-phase heat transfer [8,12–19]. A nanofluid is a suspension of some conductive powders such as metal oxides uniformly dispersed within a common cooling fluid such as water or oil [20–22]. Hence, dispersion of nanoparticles requires specific techniques such as using ultrasonic waves and controlling the pH value of the nanofluid [21].

So far, many endeavors have been made to understand the effect of nanofluids on the thermo-physical properties of nanofluids. Thermal conductivity is one of the key physical properties of the nanofluids, which strongly influences the HTC of the system. Together with thermal conductivity, the heat capacity of the working fluid is another important parameter, which can improve the thermal characteristics of a system. Studies have also shown that the presence of the nanoparticles in the liquid adds some nano-scale phenomena into the heat-transfer mechanism. The first important contributor is the Brownian motion of the nanoparticles that enhances the heat transfer by random walk and random motion of the particles within the system [23,24]. In a random movement, particles collide with the hot walls and absorb the thermal energy by conduction mechanism and carry the thermal energy to the cold region and release the thermal energy with the convection heat transfer. The thermophoresis phenomenon is another participant to the heat-transfer increment, in which the nanoparticles move from the hot wall to the cold wall due to the temperature gradient between the cold and hot wall [25].

Considering the above advantages of the nanofluids, many endeavors have been undertaken to use the nanofluids in thermal systems. The usage of the nanofluids in HPs has substantially been reviewed in the literature. For instance, Shafahi et al. [26] numerically studied the HTC of three different nanofluids including $Al_2O_3$, $TiO_2$, and CuO nanoparticles flowing through a cylindrical pipe. They modeled the system using a 2D model by assuming a steady state condition as well as incompressible and Newtonian nanofluids. Their outcomes illustrated that by using the nanofluids, thermal resistance of the system is suppressed, and the temperature gradient is reduced due to the promotion of the heat transfer. Also, using nanoparticle with smaller diameter or at higher concentration can potentially reduce the thermal resistance of the system. Naphon et al. [27] perused the influence of $TiO_2$ nanoparticle with average size of 21 nm on the heat-transfer mechanisms of alcohols in the HPs. They observed that by adding particles at 0.1% of volume fraction to the alcohol, the heat-transfer coefficient (HTC) can be enhanced by 10.6%. Azari and Derakhshandeh [28] performed some experiments to study the efficacy of aluminum oxide/ $H_2O$ nanofluid on the HTC of a HP heat exchanger in the presence of butterfly tube inserts when the HP is at constant heat flux. The results illustrated that by adding nanoparticle to the system, a significant enhancement for Nusselt number was gained (~345%). Abdollahi-Moghaddam et al. [29] carried out an experiment on CuO/ $H_2O$ nanofluid flowing through a tube to investigate the energy efficiency of a system working with HP. Also, an ANN model was expanded to investigate the thermal efficiency of the HP. It was identified the efficiency of the system can be improved with nanofluids. It was also understood that the heat transfer can be promoted up to 2.8 times, compared to the base fluid. They demonstrated that using a nanofluid can suppress the consumption of working fluid by 37% and can diminish the size of the system by 55%. Sarafraz et al. [11] carried a set of experiments out, to peruse a HP working with $ZrO_2$/acetone nanofluid. The setup was fabricated to simulate a solar collector. As expected, nanofluid diminishes the thermal resistance and gained HTC by 36.3%. Abdul Hamid et al. [6] experimentally analyzed the efficacy of $TiO_2$-$SiO_2$/water-EG nanofluid on the HTC of a HP. They found the heat-transfer efficiency can be improved by 254.4%, whereas the friction factor was also augmented by 76%, compared to the base fluid. Sarafraz et al. [11] fulfilled several experiments to study the efficacy of CNT/water nanofluid on HTC of a flat HP. They proved that the HTC of the system was improved by 40%, compared to water. HTC of pulsating HP (PHP) using 0.25–1.5 g/lit of $GO/H_2O$ nanofluids was experimentally studied by Nazari et al. [30]. Their results demonstrated that the thermal resistance of a HP can be diminished by 42%, while thermal conductivity and viscosity both increased when nanofluid was employed in the HP. Togun et al. [31] modeled a HP including a double forward-facing steps where CuO/water and $Al_2O_3$/water were employed inside the HP. The results showed that Nusselt number can be augmented by raising the nanofluid's concentration, height of the step, and the fluid velocity. Also, they found that alumina/water nanofluid at vol.% = 4 has the highest HTC.

Facing with the above literature, it can be stated that despite immense accomplished studies on the potential usage of the nanofluids in various thermal systems, less studies have been devoted to the solar thermal energy with the focus on the application of the nanofluids in smart cities, in which energy plays a key role in the municipal development. Renewable energy can open a new door towards the sustainability of the smart cities and nanofluid can contribute to this. Hence, in this study, a fundamental research is fulfilled to assess the feasible application of graphene nanoplatelets-n-pentane as the working fluid inside the HPs. N-pentane with the very high vapor pressure can be a fast-responding working fluid in the HP. Also, the presence of the graphene can improve the thermal characteristics of n-pentane. Accordingly, a thermosyphon HP is employed and effect of several operating parameters such as the applied thermal energy to the evaporator, the installation tilt angle, the filling ratio (FR) of the working fluid and the mass fraction of the nanofluid on the HTC of the evaporator and the efficiency of the HP is experimentally studied. The developed nanofluid can take advantage of good thermal properties of n-pentane while also improving the efficiency of the system due to the presence of the graphene nanoplatelets. N-pentane has reasonably good thermal features and has higher vapor pressure than alcohols and acetone. Therefore, it can easily evaporate and absorb large amount of heat due to the latent heat. However, the thermal conductivity of n-pentane is weak; thereby adding graphene can plausibly increase the thermal conductivity of the working fluid. For solar applications, high heat removal capability is a key characteristic which needs a working fluid with quick response. Due to the large vapor pressure and plausible thermal performance, n-pentane/graphene was selected for the present study.

## 2. Experimental

### 2.1. Test Rig

Figure 1 displays a schematic of the used apparatus for assessing the HTC of the HP. The test rig comprised of a charging loop, which was used to charge the working fluid into the HP. The charging loop included a vacuum pump, a syringe pump, and the circulation pipes. The volume of the working fluid charged into the system was accurately measured with the syringe pump. Using the charging loop, the value of the HTC was determined. Notably, during the charging process, the HP was de-aerated with the vacuum pump at 10 kPa. The heart of the system was a gravity-assisted thermosyphon with an evaporator with rough walls, adiabatic section in the middle and a condenser part connected to a refrigerant cooling system. The evaporator part was heated up with an flexi cartridge with power throughput of 1400 Watt. Three thermocouples were installed on the evaporator section, two on the adiabatic, and three on the condenser to record the temperature change along with the length of the HP. Also, the HP was heavily insulated to minimize the heat loss to the environment. Notably, the HP was purchased from RS-components Company and the working fluid was only replaced with the nanofluids. All the thermocouples were connected to a data logger with frequency of 1 kHz connected to a PC.

The installation angle (referred to as tilt angle) of the HP was changed with a base equipped with an inclinator to accurately measure the angle of the HP with the horizon line. The experiments were conducted for various installation angles, HTCs, heat fluxes and mass fraction of the nanoparticles and the temperature distribution profile and HTC of the evaporator were measured. The internal pressure of the HP was constantly monitored to ensure that it was not pressured during the charging process.

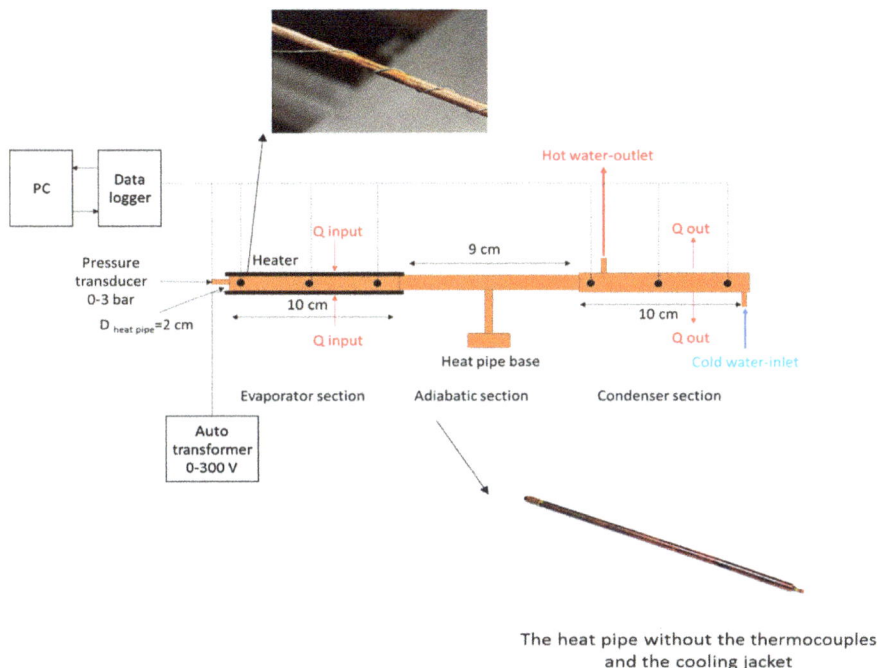

**Figure 1.** A schematic illustration of the studied apparatus.

## 2.2. Preparation of the Working Fluid

To prepare the nanofluids, graphene nanoplatelets were purchased from USNANO Company and were dispersed in n-pentane purchased from Sigma. Following method was followed to obtain the nano-suspension of graphene-n-pentane nanofluid (GNP-pentane):

(1) Initially, the graphene nanoplatelets were weighted with a balancer. For 1 kg of n-pentane, the desired mass of graphene platelets was dispersed in n-pentane such that the nanofluids were prepared at weight percentages of 0.1, 0.2 and 0.3.

(2) A magnetic stirrer was used to uniformly disperse the GNPs into n-pentane at stirring speed of 300 rpm for 10 minutes. However, due to the potential agglomeration of the nanoparticles, an ultrasonic homogenizer was employed at 350 Watt and frequency of 40 kHz for almost 10 minutes to crack the clusters and agglomerated particles.

(3) To increase the stability of the nanoparticles, an anionic surfactant of nonyl phenol ethoxylate was added to the nano-suspensions and the nano-suspensions were stirred for 5 more minutes to uniformly disperse the surfactant inside the nano-suspension.

(4) pH of the nano-suspensions was also regulated with a buffer solution to minimize the fouling formation within the system. A mixture of HCl and NaOH at 0.1 mM was employed.

(5) Time-settlement experiments were employed to ensure about the stability of the nanofluids. To perform, nanofluids were placed inside the containers at various pH values and the

sedimentation of the nanoparticles was constantly monitored to identify the best conditions in which the thickness of the sedimentation layer is minimized. Stability tests results illustrated that nanofluid could be stable for three weeks, which was sufficient for conducting the experiments.

### 2.3. Data Reduction

To measure the thermal resistance of the evaporator part, can be represented by:

$$R = \frac{T_e - T_a}{Q} \tag{1}$$

Here, $Q$ is the applied heat to the evaporator section, $T_e$ and $T_a$ are the evaporator's and adiabatic temperatures. Also, the HTC of the evaporator section was calculated with the following equation:

$$h = \frac{q''}{T_e - T_a} = \frac{V \times I}{T_e - T_a} \tag{2}$$

Here, $V$ and $I$ are the voltage and current applied to the AC flexi cartridge, which according to Joule's effect can control the applied heat to the evaporator section.

The FR of the HP is defined as follows:

$$FR = \frac{v_{nf}}{v_{evap.}} \tag{3}$$

Here, $v_{nf}$ is the volume of the working fluid in the evaporator (cm$^3$) and $v_{evap.}$ is the total volume of the evaporator (cm$^3$). The amount of FR was controlled with the syringe pump flow meter. Tilt angle is also defined as the angle between the body of the hat pipe and the horizon line and it was measured using inclinometer. To assess the uncertainty of the tests, the equation presented by Kline-McClintock [32] was used and the uncertainty of 4.5% and 5.1% were obtained for the heat-transfer coefficient and thermal resistance, respectively considering the accuracy of 1% for thermocouples, 1% for voltage and current.

To evaluate the uncertainty analysis, Kline-McClintock technique was employed [33]. Considering the accuracy of the instruments (1% of reading value for thermocouple, 0.1% for voltage and current, 1% for inclination meter and 0.1% for the syringe pump), the uncertainty value for the heat-transfer coefficient of the evaporator was ±6.9%.

## 3. Results and Discussion

### 3.1. Temperature Profile

Figure 2 displays the temperature distribution along the length of the HP for pure pentane and GNP-n-pentane nanofluid at q'' = 5 kW/m$^2$. As shown, the temperature in the evaporator is high, while gradually reduces in the adiabatic and reaches the lowest at the condenser. For pentane, the temperature of the evaporator is relatively close to the boiling temperature of pentane, hence, there is a potential for incipience of the boiling heat transfer in the evaporator section. If boiling hat transfer occurs in the evaporator, due to the formation of the bubbles, the HTC considerably increases, which can be attributed to the bubble interactions and the local agitation due to the bubble movements close to the walls of the evaporator. Interestingly, for GNP-pentane nanofluid, it can be seen that along the length of the HP, temperature decreases compared to pure n-pentane. This is because; more heat is absorbed by the working fluid due to the increase in the conduction heat transfer and thermal conductivity of the working fluid. For example, in the evaporator region, the average temperature of HP for pentane was 78 °C, while for nanofluids at wt.% = 0.1, 0.2 and 0.3, it was 75 °C, 72 °C and 69.5 °C, respectively. This is because; the GNPs absorb the thermal energy using conduction mechanism and transfer it to the cold region of the working fluid using conduction-conduction and conduction-convection mechanisms.

Hence, the heat-transfer mechanism is improved in the evaporator, which in turn improves the total HTC of the HP.

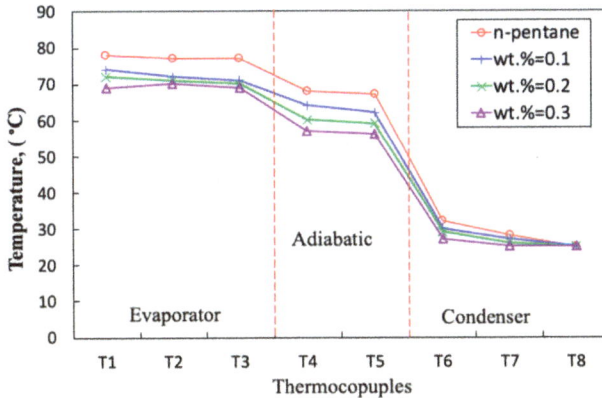

**Figure 2.** The temperature distribution of the HP for pentane and GNP-pentane nanofluids.

As can also be seen in Figure 3, the thermal resistance of the HP decreases by increasing the applied heat flux to the evaporator section. This can be ascribed to the increment in the HTC of the HP, which is an intensified at large heat flux value. Also, the thermal resistance showed an asymptotic behavior meaning that at high heat flux conditions, the thermal resistance of the HP is minimized. With adding more nanomaterials to the working fluids, the thermal resistance of the HP is more suppressed. Furthermore, the thermal resistance of the HP at heat fluxes >25 kW/m$^2$ is almost constant showing the stability of the working fluid inside the studied system.

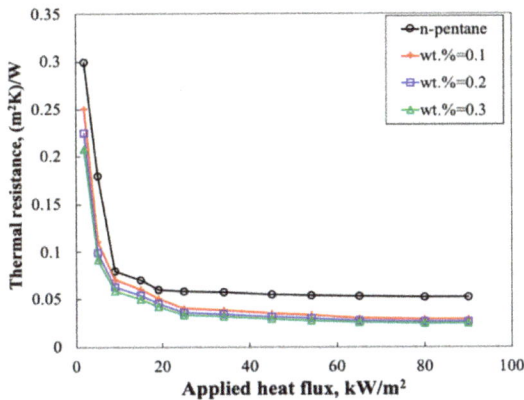

**Figure 3.** Variation of the thermal resistance with the applied heat flux for n-pentane and nanofluids.

The thermal resistance strongly depends on the temperature difference between the adiabatic and the evaporator section. Initially, for small applied heat, the convective heat transfer is the dominant mechanism of heat transfer, hence, with an increase in the applied heat, the temperature of the working fluid increases reaching the boiling point of the working fluid. Then, with an increase in the applied, only the bubble formation can potentially increase, and temperature of the working fluid remains constant. Hence, for high applied heat to the evaporator, the thermal resistance or thermal performance off the HP remains unchanged.

## 3.2. Effect of Heat Flux

Figure 4 presents the alteration of the HTC of the evaporator with the applied heat flux to the evaporator. As can be seen, with increasing the applied heat flux to the evaporator region, the HTC of the heat pipe increases, which can be attributed to the intensification of the heat-transfer mechanism from convective heat transfer towards the nucleate boiling. Also, at higher heat fluxes, the rate of the evaporation increases as well. Notably, the slope of increase from a specific heat flux changes and is promoted. For example, for heat fluxes <25 kW/m$^2$, the HTC increases slightly with heat flux; however, for heat fluxes >25 kW/m$^2$ the HTC increases significantly, which can be attributed to the incipience of the nucleate boiling in the evaporator, which promotes the HTC of the system. As an example, at heat flux (HF) of 20 kW/m$^2$, the HTC for pentane was 1755 W/(m$^2$K), though, at HF = 45 kW/m$^2$, the HTC was 2900 W/(m$^2$K). Also, for the nanofluids, the HTC was larger than those of measured for pentane. For instance, for wt. % = 0.1, at HF = 90 kW/m$^2$, the HTC was 7900 W/m$^2$. K, whereas it was 8210 W/m$^2$. K and 9820 W/m$^2$. K for wt. % = 0.2 and 0.3, respectively.

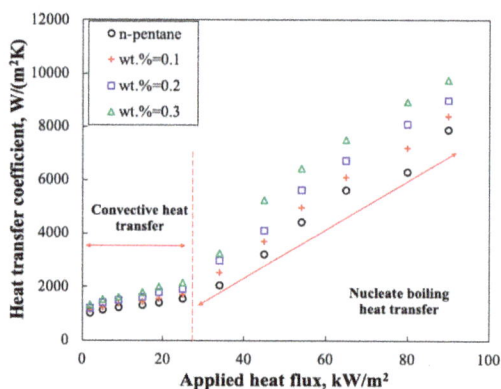

**Figure 4.** Alteration of the HTC with heat flux for n-pentane and nanofluids.

Figure 5 also shows the enhancement in the HTC of the evaporator. As can be seen, the enhancement in the HTC for region in which HF is <25 kW/m$^2$ is relatively small in comparison with those of recorded for the HF >25 kW/m$^2$. For example, for wt.% = 0.1, the HTC enhancement at HF of 15 kW/m$^2$, and 25 kW/m$^2$ was 1.08 and 1.12, respectively. However, at HF of 35 kW/m$^2$ and 80 kW/m$^2$, it was 1.24 and 1.13. As discussed, this can be attributed to the incipience of the nucleate boiling heat transfer, which further promotes the HTC of the system. Also, owing to the presence of the GNPs, the Brownian motion and the thermophoresis effect of the GNPs, improves the HTC of the system. Hence, nanofluid showed better HTC in comparison with pure pentane.

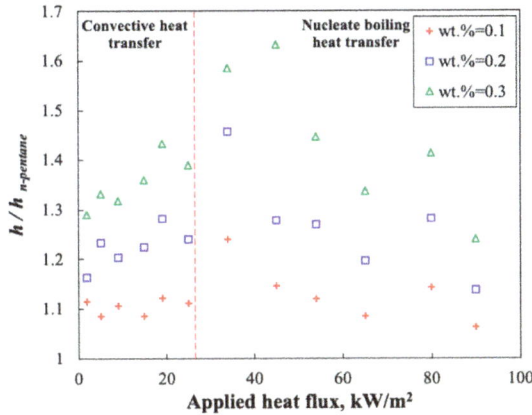

**Figure 5.** Variation of $h_{nanofluid}/h_{n-pentane}$ with the applied heat flux for various mass fractions of graphene nanoplatelets in n-pentane.

### 3.3. Filling Ratio

Figure 6 shows the variation of the HTC of the system with the HTC for heat flux of 65 kW/m$^2$ for pure pentane and nanofluids. In this heat flux, nucleate boiling occurs as temperature of the evaporator was above 100 °C. As is clear, initially, with increasing the HTC of the HP, the HTC of the system increases reaching to a point that a trade-off occurs between the space available in the evaporator for the migration of vapor to the condenser and the amount of the liquid available in the evaporator. On the other hand, with a raise in the amount of the working fluid, the thermal capacity of the fluid augments. It should be noted that the space available in the HP becomes a limiting factor. Therefore, there is an optimum for the HTC, which is 0.55 for the present work. For example, at FR = 0.55 the HTC for water and nanofluids at 0.1%, 0.2% and 0.3% are maximized which are 5200 W/(m$^2$K), 5450 W/(m$^2$K), 5510 W/(m$^2$K) and 5700 W/(m$^2$K), respectively.

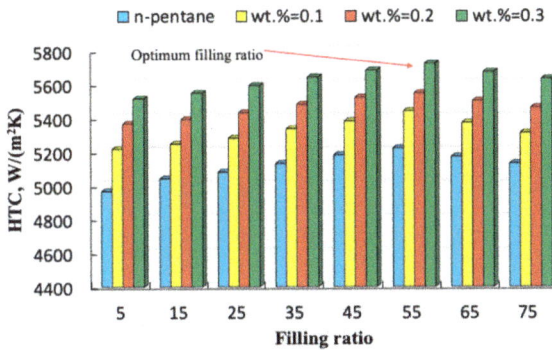

**Figure 6.** Variation of the HTC of the system with the HTC value for pure n-pentane and nanofluids.

### 3.4. Installation Angle

Figure 7 demonstrates the dependence of the HTC of the system on the installation angle (IA) of the HP for pure n-pentane and nanofluids. As can be seen, there is another trade-off identified for the HP such that with increasing the installation angle, the HTC of the system initially intensifies reaching to a point in which the HTC starts to diminish with increasing the installation angle. For example, at IA = 5, the HTC for n-pentane and nanofluid at wt.% = 0.3 was 4500 W/(m$^2$K) and 4900 W/(m$^2$K), respectively.

However, by increasing the IA to 65°, the HTC of the system improved to 4800 and 5300 W/(m²K), respectively and decreased to 4600 W/(m²K) and 5085 W/(m²K) for IA = 75°. This can be attributed to the effect of gravity forces on the liquid working fluid. For small angles, the effect of gravity is insignificant; hence return of the working fluid to the evaporator is not strongly affected by the gravity. However, for larger angles, such return is facilitated due to the gravity. However, this adds another trade-off to the system. By facilitating the return of the working fluid to the evaporator, the residence time inside the condenser considerably decreases and working fluid does not have sufficient time to transfer the heat to the surrounding. Hence, at installation angle of 65°, the largest HTC was recorded for the system. This angle was the same for nanofluids and for the n-pentane as well.

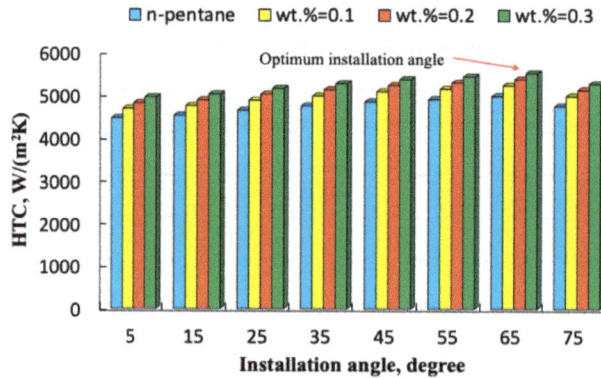

**Figure 7.** Variation of the HTC with the installation angle of the HP for pure n-pentane and nanofluids.

It is worth saying that there are several operating parameters which can potentially limit the heat-transfer rate and operation of the HP. These parameters include rate of the boiling heat transfer, the capillary effect, vapor velocity (sonic limitation) and entrainment of the liquid and vapor phases. Depending on the structure of the thermosyphon and heat pipe, any of above parameters can affect the thermal performance of the system. If the rate of the boiling heat transfer is relatively large, the dry-out phenomenon and critical heat flux occur, which can damage the evaporator section of the heat pipe. Also, if the rate of the evaporation is large, the gas velocity can reach to the sonic level (if diameter is very small, which causes a limitation in the heat transfer. Also, if the amount of vapor is high, it can causes a choked flow or even entrainment in the adiabatic or evaporator sections [34,35].

## 4. Conclusions

In this investigation, results of an experimental investigation on the HTC of a HP working with graphene nanoplatelets-n-pentane nanofluid were reported. It was identified that the presence of graphene nanoplatelets promoted the HTC of the HP. It was also found that by boosting the applied heat flux to the evaporator, the HTC of the system augmented. This was ascribed to the incipience of the nucleate boiling heat transfer of the evaporator. However, HTC and installation angle represented trade-off behavior in the system. The former showed trade-off behavior between the space available in the evaporator and the amount of the liquid working fluid; however, the latter revealed a trade-off between the effect of the gravity force on the liquid working fluid and the residence time of the working fluid inside the condenser. Overall, the presence of the graphene nanoplatelets promoted the Brownian motion and thermophoresis effect, which in turn improved the HTC of the system.

**Author Contributions:** Conceptualization, M.R.S and M.M.S.; methodology, All authors equally contributed to this; formal analysis, All authors equally contributed to this; writing—original draft preparation, M.M.S., M.G. and M.R.S; writing—review and editing, All authors equally contributed to this.

**Funding:** The authors extend their appreciation to the Deanship of Scientific Research at Majmaah University for supporting this work under project number No. (RGP-2019-17). Zhe Tian wants to acknowledge the following grants: NSFC (51709244), Taishan Scholar (tsqn201812025) and Fundamental Research for Central Universities (201941008).

**Acknowledgments:** The first and fifth authors of this work tend to appreciate Rayan Sanat Co. for their financial supports and sharing the HP facility.

**Conflicts of Interest:** The authors declare no conflict of interest.

## References

1. Tsai, C.; Chien, H.; Ding, P.; Chan, B.; Luh, T.-Y.; Chen, P.-H. Effect of structural character of gold nanoparticles in nanofluid on heat pipe thermal performance. *Mater. Lett.* **2004**, *58*, 1461–1465. [CrossRef]
2. Panwar, N.; Kaushik, S.; Kothari, S.; Panwar, D.N.L. Role of renewable energy sources in environmental protection: A review. *Renew. Sustain. Energy Rev.* **2011**, *15*, 1513–1524. [CrossRef]
3. Dincer, I. Renewable energy and sustainable development: A crucial review. *Renew. Sustain. Energy Rev.* **2000**, *4*, 157–175. [CrossRef]
4. Kreith, F.; Kreider, J.F. *Principles of Solar Engineering*; Hemisphere Publishing Corp.: Washington, DC, USA, 1978; p. 790.
5. Tian, Y.; Zhao, C. A review of solar collectors and thermal energy storage in solar thermal applications. *Appl. Energy* **2013**, *104*, 538–553. [CrossRef]
6. Olia, H.; Torabi, M.; Bahiraei, M.; Ahmadi, M.H.; Goodarzi, M.; Safaei, M.R. Application of Nanofluids in Thermal Performance Enhancement of Parabolic Trough Solar Collector: State-of-the-Art. *Appl. Sci.* **2019**, *9*, 463. [CrossRef]
7. Kreider, J.F.; Kreith, F. *Solar Energy Handbook*; McGraw-Hill: New York, NY, USA, 1981.
8. Arya, A.; Sarafraz, M.M.; Shahmiri, S.; Madani, S.A.H.; Nikkhah, V.; Nakhjavani, S.M. Thermal performance analysis of a flat heat pipe working with carbon nanotube-water nanofluid for cooling of a high heat flux heater. *Heat Mass Transf.* **2018**, *54*, 985–997. [CrossRef]
9. Sarafraz, M.; Hormozi, F.; Peyghambarzadeh, S. Thermal performance and efficiency of a thermosyphon heat pipe working with a biologically ecofriendly nanofluid. *Int. Commun. Heat Mass Transf.* **2014**, *57*, 297–303. [CrossRef]
10. Sarafraz, M.; Hormozi, F. Experimental study on the thermal performance and efficiency of a copper made thermosyphon heat pipe charged with alumina—Glycol based nanofluids. *Powder Technol.* **2014**, *266*, 378–387. [CrossRef]
11. Sarafraz, M.; Pourmehran, O.; Yang, B.; Arjomandi, M. Assessment of the thermal performance of a thermosyphon heat pipe using zirconia-acetone nanofluids. *Renew. Energy* **2019**, *136*, 884–895. [CrossRef]
12. Nakhjavani, M.; Nikkhah, V.; Sarafraz, M.; Shoja, S.; Sarafraz, M. Green synthesis of silver nanoparticles using green tea leaves: Experimental study on the morphological, rheological and antibacterial behaviour. *Heat Mass Transf.* **2017**, *53*, 3201–3209. [CrossRef]
13. Nikkhah, V.; Sarafraz, M.; Hormozi, F. Application of spherical copper oxide (II) water nano-fluid as a potential coolant in a boiling annular heat exchanger. *Chem. Biochem. Eng. Q.* **2015**, *29*, 405–415. [CrossRef]
14. Sarafraz, M.; Hormozi, F.; Peyghambarzadeh, S.; Vaeli, N. Upward Flow Boiling to DI-Water and Cuo Nanofluids Inside the Concentric Annuli. *J. Appl. Fluid Mech.* **2015**, *8*, 651–659.
15. Sarafraz, M.; Hormozi, F.; Silakhori, M.; Peyghambarzadeh, S.M. On the fouling formation of functionalized and non-functionalized carbon nanotube nano-fluids under pool boiling condition. *Appl. Therm. Eng.* **2016**, *95*, 433–444. [CrossRef]
16. Sarafraz, M.; Nikkhah, V.; Madani, S.; Jafarian, M.; Hormozi, F. Low-frequency vibration for fouling mitigation and intensification of thermal performance of a plate heat exchanger working with CuO/water nanofluid. *Appl. Therm. Eng.* **2017**, *121*, 388–399. [CrossRef]
17. Sarafraz, M.; Nikkhah, V.; Nakhjavani, M.; Arya, A. Thermal performance of a heat sink microchannel working with biologically produced silver-water nanofluid: Experimental assessment. *Exp. Therm. Fluid Sci.* **2018**, *91*, 509–519. [CrossRef]

18. Sarafraz, M.M.; Peyghambarzadeh, S.; Fazel, S.A.; Vaeli, N. Nucleate pool boiling heat transfer of binary nano mixtures under atmospheric pressure around a smooth horizontal cylinder. *Period. Polytech. Chem. Eng.* **2013**, *57*, 71–77. [CrossRef]

19. Sarafraz, M.; Peyghambarzadeh, S.M. Nucleate pool boiling heat transfer to Al$_2$O$_3$-water and TiO$_2$-water nanofluids on horizontal smooth tubes with dissimilar homogeneous materials. *Chem. Biochem. Eng. Q.* **2012**, *26*, 199–206.

20. Yu, W.; France, D.M.; Routbort, J.L.; Choi, S.U.S. Review and comparison of nanofluid thermal conductivity and heat transfer enhancements. *Heat Transf. Eng.* **2008**, *29*, 432–460. [CrossRef]

21. Li, Y.; Zhou, J.; Tung, S.; Schneider, E.; Xi, S. A review on development of nanofluid preparation and characterization. *Powder Technol.* **2009**, *196*, 89–101. [CrossRef]

22. Kleinstreuer, C.; Feng, Y. Experimental and theoretical studies of nanofluid thermal conductivity enhancement: A review. *Nanoscale Res. Lett.* **2011**, *6*, 229. [CrossRef]

23. Jang, S.P.; Choi, S.U.S. Role of Brownian motion in the enhanced thermal conductivity of nanofluids. *Appl. Phys. Lett.* **2004**, *84*, 4316–4318. [CrossRef]

24. Prasher, R.; Bhattacharya, P.; Phelan, P.E. Brownian-motion-based convective-conductive model for the effective thermal conductivity of nanofluids. *J. Heat Transf.* **2006**, *128*, 588–595. [CrossRef]

25. Malvandi, A.; Ganji, D.D. Brownian motion and thermophoresis effects on slip flow of alumina/water nanofluid inside a circular microchannel in the presence of a magnetic field. *Int. J. Therm. Sci.* **2014**, *84*, 196–206. [CrossRef]

26. Shafahi, M.; Bianco, V.; Vafai, K.; Manca, O. Thermal performance of flat-shaped heat pipes using nanofluids. *Int. J. Heat Mass Transf.* **2010**, *53*, 1438–1445. [CrossRef]

27. Naphon, P.; Khonseur, O. Study on the convective heat transfer and pressure drop in the micro-channel heat sink. *Int. Commun. Heat Mass Transf.* **2009**, *36*, 39–44. [CrossRef]

28. Azari, A.; Derakhshandeh, M. An experimental comparison of convective heat transfer and friction factor of Al$_2$O$_3$ nanofluids in a tube with and without butterfly tube inserts. *J. Taiwan Inst. Chem. Eng.* **2015**, *52*, 31–39. [CrossRef]

29. Abdollahi-Moghaddam, M.; Motahari, K.; Rezaei, A.; Abdollahi-Moghaddam, M.; Motahari, K.; Rezaei, A. Performance characteristics of low concentrations of CuO/water nanofluids flowing through horizontal tube for energy efficiency purposes; an experimental study and ANN modeling. *J. Mol. Liq.* **2018**, *271*, 342–352. [CrossRef]

30. Nazari, M.A.; Ghasempour, R.; Ahmadi, M.H.; Heydarian, G.; Shafii, M.B. Experimental investigation of graphene oxide nanofluid on heat transfer enhancement of pulsating heat pipe. *Int. Commun. Heat Mass Transf.* **2018**, *91*, 90–94. [CrossRef]

31. Togun, H.; Ahmadi, G.; Abdulrazzaq, T.; Shkarah, A.J.; Kazi, S.N.; Badarudin, A.; Safaei, M.R. Thermal performance of nanofluid in ducts with double forward-facing steps. *J. Taiwan Inst. Chem. Eng.* **2015**, *47*, 28–42. [CrossRef]

32. Kline, S.J.; McClintock, F.A. Describing Uncertainties in Single-Sample Experiments. *Mech. Eng.* **1953**, *75*, 3–8.

33. Kline, S. The purposes of uncertainty analysis. *J. Fluids Eng.* **1985**, *107*, 153–160. [CrossRef]

34. Faghri, A. Heat pipes: Review, opportunities and challenges. *Front. Heat Pipes* **2014**, *5*. [CrossRef]

35. Naphon, P.; Thongkum, D.; Assadamongkol, P. Heat pipe efficiency enhancement with refrigerant—Nanoparticles mixtures. *Energy Convers. Manag.* **2009**, *50*, 772–776. [CrossRef]

*applied sciences*

MDPI

*Article*

# Potential of Solar Collectors for Clean Thermal Energy Production in Smart Cities using Nanofluids: Experimental Assessment and Efficiency Improvement

**M. M. Sarafraz [1], Iskander Tlili [2], Mohammad Abdul Baseer [3] and Mohammad Reza Safaei [4,5,\***

[1]  School of Mechanical Engineering, the University of Adelaide, Adelaide, South Australia, Australia; mohammadmohsen.sarafraz@adelaide.edu.au

[2]  Department of Mechanical and Industrial Engineering, College of Engineering, Majmaah University, Al-Majmaah 11952, Saudi Arabia; l.tlili@mu.edu.sa

[3]  Department of Electrical Engineering, College of Engineering, Majmaah University, Al-Majmaah 11952, Saudi Arabia; m.abdulbaseer@mu.edu.sa

[4]  Division of Computational Physics, Institute for Computational Science, Ton Duc Thang University, Ho Chi Minh City, Vietnam

[5]  Faculty of Electrical and Electronics Engineering, Ton Duc Thang University, Ho Chi Minh City, Vietnam

\*   Correspondence: cfd_safaei@tdtu.edu.vn; Tel.: +15-026579981

Received: 3 April 2019; Accepted: 27 April 2019; Published: 7 May 2019

**Abstract:** In this article, an experimental study was performed to assess the potential thermal application of a new nanofluid comprising carbon nanoparticles dispersed in acetone inside an evacuated tube solar thermal collector. The effect of various parameters including the circulating volumetric flow of the collector, mass fraction of the nanoparticles, the solar irradiance, the tilt angle and the filling ratio values of the heat pipes on the thermal performance of the solar collector was investigated. It was found that with an increase in the flow rate of the working fluid within the system, the thermal efficiency of the system was improved. Additionally, the highest thermal performance and the highest temperature difference between the inlet and the outlet ports of the collector were achieved for the nanofluid at wt. % = 0.1. The best tilt angle and the filling ratio values of the collector were 30° and 60% and the maximum thermal efficiency of the collector was 91% for a nanofluid at wt. % = 0.1 and flow rate of 3 L/min.

**Keywords:** solar collector; evacuated tube; carbon-acetone nanofluid; thermal performance

## 1. Introduction

Solar thermal energy is a free and abundant source of energy. However, the intermittent access to the solar thermal energy during the day and also during the off-sun period together with a lack of reliable and mature technology are drawbacks of using solar thermal energy [1–3]. Hence, immense research has been done to utilize solar thermal energy in industrial and domestic sectors.

Solar thermal collectors are devices that absorb solar thermal energy and an evacuated solar thermal collector is an apparatus with several evacuated tubes embedded with a heat pipe (HP), which absorbs the solar irradiance and converts it into thermal energy using phase change mechanism inside the heat pipes. Each evacuated tube has an internal HP, which delivers the heat from the absorbance to the header tank. A HP is a two-phase heat exchanging medium, consisting of three domains of evaporator, adiabatic and the condenser sections. Generally, the condenser section of the HP is constantly cooled with water, while the evaporator section is exposed to the high heat flux thermal energy absorbed from the sun [4,5]. Hence, in the evaporator section, boiling is the general heat transfer mechanism, which provides a large heat transfer coefficient (HTC) [6–8]. This is because the generation of the bubbles in the evaporator section not only improves the heat transfer in the evaporator but also provides an efficient phase change and two-phase heat transfer throughout the HP.

The working fluid inside the HP is a key parameter determining the overall thermal performance (TP) of the system. A working fluid with a low boiling temperature, high thermal conductivity and plausible thermal properties can increase the TP of the HP leading to more heat transfer within the collector. Since the introduction of nanofluid by the Aragon National Laboratory (ANL) [9–11], much effort has been made to utilize it in various thermal applications including solar thermal energy [12–25]. For example, Dehaj et al. [26] performed experiments to study the TP of an evacuated HP solar collector. They used nanofluids to boost the heat transfer within the solar collector. The influence of the fluid flow rate of the MgO/water nanofluid within the collector together with the particle's concentration on the efficiency of the collector was experimentally investigated and it was found that the efficiency of the collector enhanced by increasing the flow rate. In addition, the efficiency of the solar collector was further improved with augmentation of MgO based nanofluid concentration. In another study, Ziyadanogullari et al. [27] experimentally assessed the influence of the presence of various nanoparticles on the TP of a flat plate collector. Nanoparticles including alumina, copper oxide and Titania were dispersed in water at volumetric concentrations of 0.2–0.8% and the flow rate of the carrying fluid was set to 250 L/h for each nanofluid. It was found that the use of nanofluid increased the collector efficiency in comparison with water. Sharafeldin et al. [28] performed experiments to quantify the TP of the evacuated tube solar collector (ETSC) working with $WO_3$ suspended in water at vol.% of 0.014, 0.028, and 0.042. They demonstrated that the TP of the collector can be improved by 21% for the nanofluid. Also, the efficiency of the system increased by 23% in comparison with the system without nanofluid. With nanofluid inside the collector, the maximum thermal efficiency of 72.8% was obtained. Cesium oxide was also identified as a plausible nanoparticle and was tested in an ETSC [29]. Using zeta potential value, they stabilized the nanofluid and the TP of the ETSC was quantified. It was demonstrated that the TP of the collector can be improved for the case of the nanofluid. The maximum efficiency of 34% was recorded for the system.

Ozsoy et al. [30] conducted a set of experiments to quantify the TP of a thermosyphon HP with a potential to be used in an ETSC. Silver-water nanofluid was used in the system with the view to evaluate its potential application for a continuous operation in the system. The nanoparticles were synthesized to ensure that the particles have longer stability. They found that the nanofluid enhanced the efficiency of solar collector between 20.7–40% in comparison with pure water and it was also identified that the use of silver-water nanofluid can provide better heat transfer and TP within the system. To further intensify the influence of nanoparticles on the thermo-hydraulic performance of the system, several modifications have been suggested by the authors. The most promising one is the functionalization of the nanoparticles with the view to increase their stability, thermal conductivity and other thermo-physical properties of the nanofluids.

Heat pipes and thermosyphons are the heart of the solar collectors. Hence, improving the TP of the HP influences the TP of the collectors. Therefore, extensive research has been conducted to better understand the effect of nanoparticles on the TP of the heat pipes. For example, in a study conducted by Nazari et al. [31] the thermal characteristic of a HP working with the graphene oxide nanofluid was experimentally studied and it was identified that the nanofluid can amend the thermo-physical properties of the working fluid including the thermal conductivity. This also led to a 43% decrease in overall TP of the system. Arya et al. [32] studied the TP of a HP working with carbon nanotube dispersed in deionized water and showed that the TP of a HP can be improved with an increment in the value of the input heat applied to the evaporator section. In addition, they identified a trade-off behavior between the filling ratio, tilt angle and the TP of the HP. Their experiments showed that the optimum value for the filling ratio for the test HP was 0.8. In addition, the nano-suspension improved the TP of the HP by 40%. The same results were also obtained by Vijayakumar et al. [33] by investigating the potential application of copper-water and alumina-water nanofluid inside a HP. They also observed that for the case of nanofluid, the temperature profile of the HP was augmented and the thermal resistance of the HP was detracted.

For most of the works discussed in the aforementioned literature, the base fluid is normally water or a material with a boiling point larger than water. Hence, the evaporation of the working fluid within the HP at low temperature is relatively low and the dominant mechanism in the evaporator is convective mixed with the evaporation. Hence, the order of magnitude for the TP in the evaporator and in the condenser is relatively smaller than those reported for the heat pipes working in the nucleate boiling two-phase regime. Thus, in the present work, acetone as a traditional working fluid for the low-temperature heat pipes was fortified with carbon nanoparticles to be used in the ETSC. By doing this, the efficiency of the HP is expected to be higher as the boiling temperature of acetone is smaller than conventional coolants including water, ethylene glycol and oils. In addition, the carbon nanoparticles are cheap, available with a relatively good thermal conductivity. Thereby dispersing it into acetone can provide the plausible TP for the evacuated tube. The effect of different operating parameters including the mass concentration of carbon in acetone, the filling ratio, and tilt angle of the collector, operating time and the fluid flow rate within the collector on the TP of the collector was investigated.

## 2. Experimental

### 2.1. Preparation, Quality Tests and Physical Properties

Carbon nano-powder (CNP hereafter, purchased from Sigma) was used. Acetone was selected as a base fluid and was purchased from Sigma Aldrich and the characteristic tests were performed on the nanoparticles including the nitrogen adsorption test, the x-ray diffraction, pore-particle size, and the morphology assessment with the view to further identify the quality of the carbon particles used in the present research. The morphology of the nanoparticles not only influences the particle–particle interaction but also changes the stability of the nanofluids as well. Also, the composition of the carbon can directly influence the thermal conductivity together with other thermo-physical properties of the nanofluid. Notably, most of the properties of the nanofluid are functions of the particle size, and the quality of the dispersion. Hence, the physical properties were experimentally measured. For nanofluids preparation, the following steps were performed:

1. Disperse the desire mass of carbon nanoparticles into acetone using magnetic stirrer at 250 rpm;
2. Add an anionic surfactant, Nonyl Phenol Ethoxilate (NPE) (at vol.% = 0.1 of total prepared nanofluid);
3. Set pH of the nanofluid to a value, in which the highest zeta potential was observed. After setting the pH value of each sample, to measure the zeta potential, the sample was assessed with a zeta sizer digital light scattering device manufactured by Malvern CO (Malvern, UK). So, the measurement was carried out after the pH was regulated using the buffer solution.
4. Sonicate the nanofluid to disperse the particles uniformly; inside the working fluid.

Figure 1 shows the image of the prepared nanofluid.

### 2.1.1. Surface Area and Particle Size

Nitrogen adsorption test was utilized to identify the average surface area and the distribution of size of the particles of the carbon nanoparticles, and the result has been represented in Figure 2. The higher surface area, the higher the TP. Prior to the adsorption experiments, samples were degassed at 140 °C for 10 h under vacuum pressure of $10^{-5}$ torr. Nitrogen adsorption isotherms were plotted using a Tristar instrument (Micromeritics, Norcross, Georgia, GA, USA) at liquid nitrogen temperature ($\sim$−196 °C). An ultra-high purity nitrogen and helium (>99.999 %) were used for the adsorption and the dead volume measurements, respectively. The Brunauer–Emmett–Teller (BET) method was employed for specific surface area evaluation [34]. The quenched solid density functional theory method

(QSDFT) [35], was utilized to evaluate the particle size diameter, (PSD) for the CNP sample. This figure exhibits a Type II isotherm based on International Union of Pure and Applied Chemistry, IUPAC classification representing the presence of mesoporous structure with negligible contributions from microspores. It also shows a very narrow Type H3 hysteresis loop, representing possible aggregates particles. The calculated BET equivalent surface area and total pore volume of the sample are relatively small (8.8 m$^3$/g and 0.027 cm$^3$/g, respectively), indicating small porosity of the sample. Total pore volume ($V_P$) was calculated based on nitrogen amount adsorbed at near saturation pressure converted to liquid volume. The PSD for the CNP sample is shown in Figure 3. The distribution is multimodal, showing a broad range of 4–50 nm. The intensity of the distribution curve is very small. This low intensity and broad distribution of pores confirm that the available surface area is primarily due to the particle's external surface rather than its internal porosity.

**Figure 1.** Preparation and dispersion of the carbon nanoparticles in acetone in four steps.

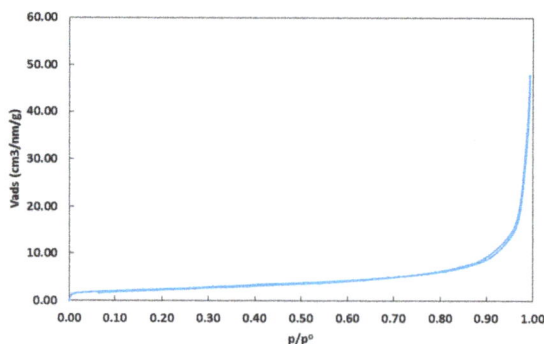

**Figure 2.** Nitrogen adsorption isotherm obtained for the carbon nanoparticle sample.

Figure 4 shows the results of particle size test for the carbon nanoparticles. Analysis of the particle size test showed that the average particle size for the carbon nanoparticles dispersed in acetone was 50 nm. This also further confirms the size claimed by the manufacturer. This also minimizes the possibility of the deposition and scale formation within the solar collector as well.

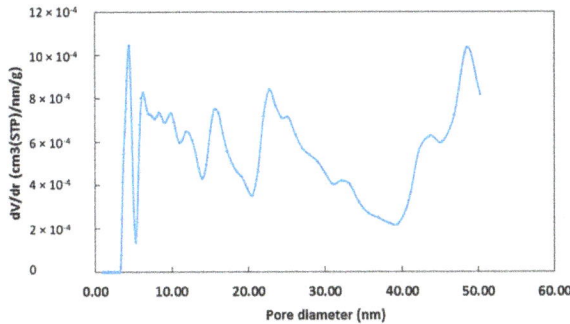

**Figure 3.** The pore size distributions obtained with quenched solid-state functional theory method for the CNP sample used in the present research.

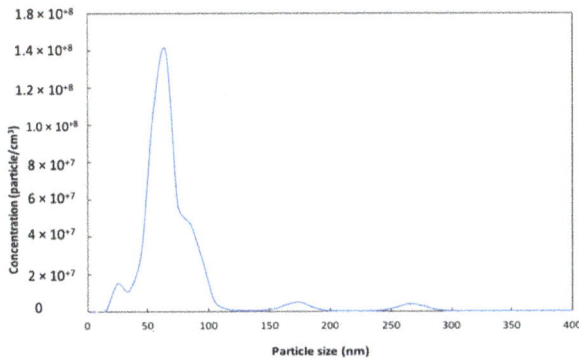

**Figure 4.** Results of particle size test for CNPs.

### 2.1.2. Structure and Composition

The x-ray diffraction test, known as XRD, was carried out for a sample of CNP using GNR APD 2000 (Italy) at ambient temperature. The data were collected in continuous mode and with the step size of $0.01^0$ and at the range of 3–80° similar to our previous works [36–39]. For the particle size determination, solid samples were dispersed in Milli-Q water in an ultrasound bath (Sonorex Super, Bandelin, Germany) for 30 min. High resolution particle detecting and sizing experiment on the CNP sample was performed using a NanoSight LM10 instrument (Malvern Instruments, Malvern, UK). Results of the XRD test have been represented in Figure 5. The peak observed at 26 degree is the main characteristic peak of carbon due to structure of the particle, which is (002) lattice plane. In addition, the small peak seen in 45 degree is due to the presence of some amorphous carbon in the system, which can be ignored.

### 2.1.3. Morphology

To assess the morphology, dispersion and agglomeration, Scanning Electron Microscopic image (SEM, represented in Figure 6a) and Transmission Electron Microscopic images, (TEM) were captured from a sample of 1% by volume of carbon nanoparticles inside the acetone as presented in Figure 6b. The required information was added to the paper. To prepare the sample for SEM imaging, the desired mass of CNP was weighted (1.25 g) and the sample was washed with methanol and then dried to remove any dust and oily material from the sample. Then the sample was coated with a conductive material using a use of a sputter-coater. To prepare the sample for TEM imaging, the prepared nanofluid was placed on the specimen grid using Formvar with a uniform thickness. The temperature of the room must be well below the boiling temperature of the liquid to prevent from the evaporation. In addition, the specimen grid disk was washed with methanol and dried to ensure that there was no impurity stick to the mesh and surface of the specimen disk. As can be seen, the morphology is spherical and the same for the particles. Importantly, neither cluster nor agglomeration is formed during the dispersion.

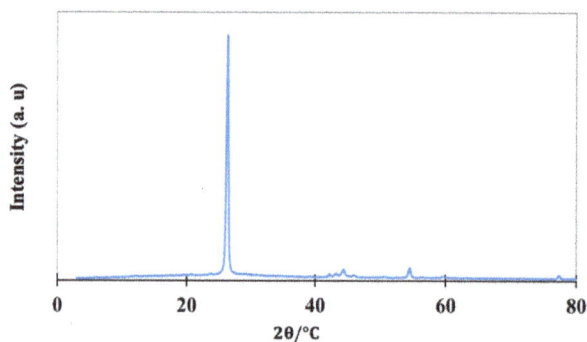

**Figure 5.** Results of XRD test conducted on carbon nanoparticles.

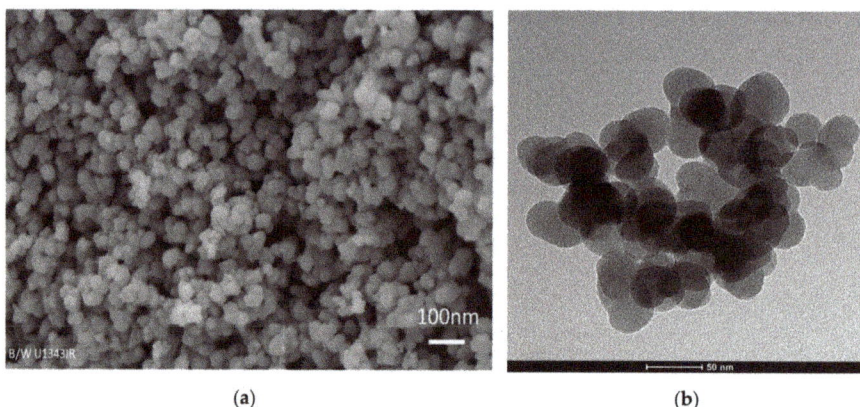

|      (a)      |      (b)      |

**Figure 6.** (a) Scanning electron microscopic image of the CNPs, (b) TEM image of CNPs.

2.1.4. Thermo-Physical Properties

To identify the influence of the nanoparticles on the thermo-physical properties of the base fluid, a series of experiments were performed with instruments to measure the thermo-physical properties of the nanofluids including the density, the thermal conductivity and the heat capacity of the nanofluid. To measure the thermal conductivity and the heat capacity of the system, a KD2 pro manufactured by Decagon (with the accuracy of ~5% and ~8% of the reading value were used. To measure the viscosity, the viscometer manufactured by Brookfield DV I series with accuracy of ~6% was used. As can also be seen in Figure 7, adding more nanoparticles to the carrying fluid, thermal conductivity of the fluid increases. For example, at 40 °C, at wt. % = 0.025, the measured thermal conductivity was ~0.21 W/(mK), while for the same temperature, at wt. % = 0.1, the thermal conductivity increased to ~0.25 W/(mK), increasing by 19%. Notably, the thermal conductivity of acetone was 0.18 W/(mK), while those of CNP-acetone were larger than pure acetone. Interestingly, with temperature increment, the thermal conductivity of the nanofluid increased. This can be attributed to the Brownian motion boosting within the bulk of the nanofluid, which enhances the collision and the conduction–conduction heat transfer between the particles. The maximum thermal conductivity was 0.27 W/(mK) at the largest temperature of 50 °C and wt. % = 0.1.

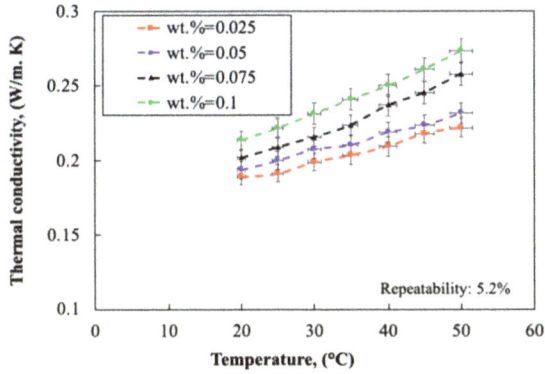

**Figure 7.** Variation of the thermal conductivity of the prepared nanofluids on temperature for various mass fractions of CNPs in acetone.

Figure 8 represents the influence of temperature on the viscosity of CNP-acetone nanofluid for different mass fractions of CNP. According to the figure, an increment in the mass fraction of the nanofluid increases the frictional forces between the layers of the base fluid increases resulting in higher viscosity of the nanofluid. In addition, temperature slightly reduces the viscosity of the nanofluid. The highest viscosity was 0.00048 Pa.s and belonged to the nanofluid at wt. % = 0.1 and T = 20 °C.

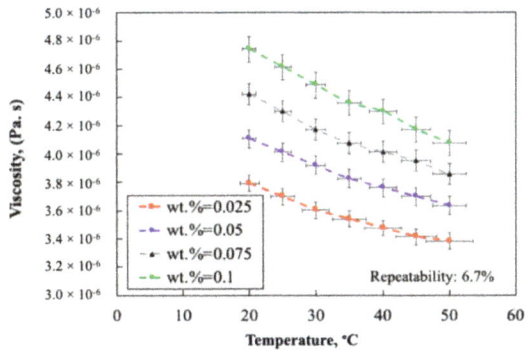

**Figure 8.** Dependence of viscosity of nanofluids on temperature for different concentrations of CNPs in acetone.

Figure 9 shows the impact of the temperature on the heat capacity of the nanofluid for various mass fractions of CNP in acetone. With temperature increment, the heat capacity boosts which is owing to the augmentation in the heat capacity of acetone. In addition, with growth in the number of nanoparticles dispersed within the nanofluid, the heat capacity of the nanofluid increases. This is because the heat capacity of carbon by itself is larger than acetone. Thereby, raising the mass fraction of the nanofluid promotes the heat capacity of the nanofluid.

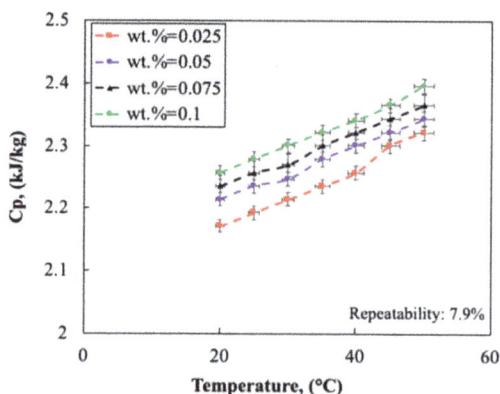

**Figure 9.** Dependence of heat capacity of nanofluids on temperature for various concentrations of CNPs in acetone.

*2.2. Experimental Setup*

2.2.1. Experimental Setup

Figure 10a shows a schematic illustration of the ET used in the solar collector, consisting of an evaporator, an adiabatic and a condenser located at the header tank. The solar irradiance is absorbed with the absorber layer and is transferred to the evaporator section in which the internal carrying fluid of the evacuated tube is evaporated with the boiling mechanism and the heat together with the vapor are transported to the header tank in which the thermal energy is delivered to the cold water. The condensed carrying fluid is returned to the bottom of the HP under a gravity-assisted falling film of liquid. The detailed information about the HP and the evacuated tube are represented in Figure 10a. Figure 10b shows the schematic of the apparatus; accompanied with an image of the ETSC. The heat pipes inside the evacuated tubes are charged with the working fluid (CNP/acetone nanofluid) and water is pumped from a reservoir to a solar collector with a centrifugal pump (manufactured by DAB). The flow rate of the flow is controlled with an ultrasonic flow meter connected to a valve and a servo. The temperature of the tank was monitored with a PID controller and a thermostat bath to ensure a back-to-back comparison between the data obtained for each test. The inlet and the outlet temperature of the solar collector was constantly monitored with two k-type thermocouples connected to a digital data logging system with reading frequency of 1 kHz. The outlet of the collector was chilled with a water jacket connected to the thermostat bath to ensure that the temperature of the tank is constant. Notably, a bypass was also utilized to adjust the flow rate within the system. This ensures a very accurate flow rate adjustment within the system.

The experiments were conducted in a geographical location with the coordination of 27.19 latitude and 56.28 longitude and situated at an elevation of 8.9 m above sea level. The location has a population of 352,173. Hence, the importance of using solar thermal energy in this city is pivotal for domestic and industrial sectors. Given the fact that this city can take advantage of almost 300 sunny days within a year, the experiments were conducted there to ensure that the maximum solar share can be achieved. The maximum solar irradiance obtained in the city was 920 W/m$^2$ well above the average of the other large cities all over the world (850–890 W/m$^2$).

(a)

(b)

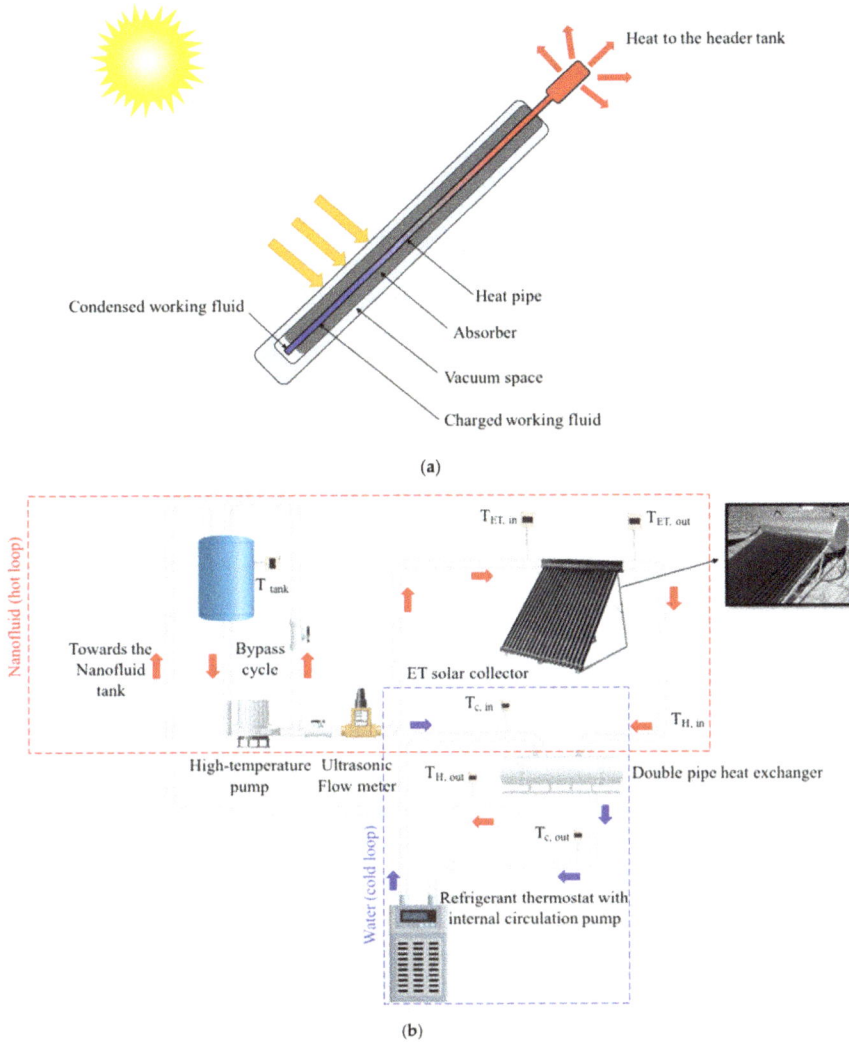

**Figure 10.** Schematic diagram of the test facility used in the present study, (**a**) the structure of the evacuated tube utilized in the solar collector, (**b**) the test rig [40]. Figure is reprinted with permission from publisher.

### 2.2.2. Data Reduction and Uncertainty Analysis

To measure the mass concentration of the nanofluid, Equation (1) was employed:

$$wt.\% = \frac{w_{np}}{w_{bf}} \tag{1}$$

Here, $w_{np}$ and $w_{bf}$ stand for the weight of carbon particles and pure acetone, respectively. In addition, the filling ratio is the volumetric ratio of a nanofluid to the total volume of the HP as follows:

$$illing\ ratio(\%) = \frac{vol.\ of\ NF}{vol.\ of\ HP} \tag{2}$$

To measure the cumulative thermal energy absorbed by the collector, at each hour the thermal energy absorbed by the collector was measured and the following correlation was used to obtain the daily cumulative thermal energy absorbed by the collector:

$$Q_{cum.} = \int_{t=1\ hr}^{t=24\ hr} Q_{col.}.dt \tag{3}$$

and

$$Q_{col.} = \dot{m} \times C_p \times (T_{ET,out} - T_{ET,in}) \tag{4}$$

Here, $C_p$ is the fluid's heat capacity, $ET, in$ and $ET, out$, stand for the inlet and outlet temperatures of the collector. In addition, $\dot{m}$ is the mass flow rate of the fluid introducing to the collector which can be calculated with Equation (5):

$$\dot{m} = \rho \times \dot{Q} \tag{5}$$

Here, $\rho$ is the density and $\dot{Q}$ is the volumetric flow rate of fluid, which can be acquired from the ultrasonic flow meter. Thus, the thermal efficiency of the collector can be obtained with the following equation:

$$\eta = \frac{\dot{m} \times C_p \times (T_{ET,out} - T_{ET,in})}{G \times A_c} \tag{6}$$

Here, G stands for the solar irradiance and $A_c$ is the area of the collector. To compute the overall value of the uncertainty of the experiments, the Moffat method was employed [41] using the values given in Table 1. According to the uncertainty values of the instruments utilized in the research, the uncertainty for the efficiency measurements is deviation within ±3.5% of the results. Table 2 also shows the geometrical specification of the evacuated tube solar collector.

**Table 1.** The accuracy and uncertainty value for the apparatuses used in this work.

| Parameter | Instrument | Uncertainty |
|-----------|------------|-------------|
| G | Pyranometer, EKO | ±1% of reading value (W/m$^2$) |
| Temperature | Omega k type thermocouple | ±1 (°C) |
| Ambient temperature | RTD, PT-100 Omega | ±0.5 (°C) |
| Flow rate | FLownetix ultrasonic sensor | 1% of reading value (L/min) |
| Tilt angle | Europac$^{TM}$ Inclinometer | ±0.1 (degree) |
| Weight | A&D balancer | ±1% of reading value (g) |

The accuracy and uncertainty value for the apparatuses used in this work.

**Table 2.** Geometrical details of the evacuated tube and the collector used in the present research.

| Parameter | Value |
|-----------|-------|
| Tube's gross area | 1.75 m$^2$ |
| Tube's aperture area | 0.77 m$^2$ |
| Total liquid capacity of the tube | 0.4 lit |
| Absorbance | 0.91 |
| Emission parameter | 0.04 |
| Collector's dimension (L × H × W) | 2 × 0.15 × 0.9 m |

## 3. Results and Discussion

### 3.1. Solar Irradiance and Ambient Temperature

Figure 11 represents the dependence on time of the solar irradiance and the hourly average ambient temperature of the location of the experiments. As can be seen, the highest solar irradiance is observed between 10:00 am and 3:00 pm reaching 920 W/m$^2$ at 1:00 pm. The solar irradiance represented in Figure 11 was obtained at the hottest day in midsummer to maximize the solar thermal energy and the thermal absorption by the solar collector. The same trend was also observed for other days in summer and other seasons, however, the highest solar irradiance value varied from 410 W/m$^2$ in winter to 520 W/m$^2$ and 780 W/m$^2$ in winter, autumn and spring, respectively. As can also be seen, the hourly mean ambient temperature followed the trend obtained for the solar irradiance such that the highest ambient temperature was obtained between 10:00 am and 3:00 pm. The lowest and the highest temperature readings were 27.5 °C and 41 °C, respectively. In addition, the temperature rectilinearly increased with the time spanning. Hence, the experiments were conducted from 6:00 am to 8:00 pm to maximize the utilization of the available solar irradiance in the region. Notably, the solar irradiance was measured with pyranometer manufactured by EKO CO. at various time of the day. The pyranometer was used to measure the planar solar irradiance on the location on which the solar collector was installed. The maximum of 920 W/m$^2$ was measured during hot days of summer in the location.

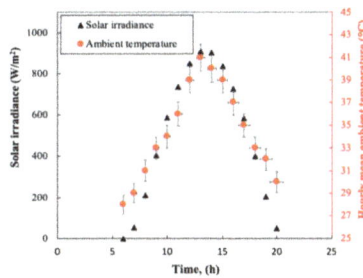

**Figure 11.** Variation of the solar irradiance and hourly ambient temperature with time.

### 3.2. Tilt Angle and Filling Ratio

Figure 12a,b represent the variation of the cumulative energy absorption from the collector with the tilt angle and filling ratio of the evacuated tubes in the 15th and 16th August, 2017 with the same solar irradiance profiles. As can be seen in Figure 12a, with an increment in the tilt angle of the collector, the cumulative energy absorption initially increased and then was suppressed with boosting the tilt angle of the collector. The highest cumulative thermal energy absorption belonged to the collector at tilt angle of 30° (the blue rectangular), while the lowest thermal energy was absorbed with the collector at tilt angle of 45° and higher. As can be seen in Figure 12b, with a growth in the filling ratio of the HP from 10% to 60%, the cumulative thermal energy absorbed by the collector increased from 14.2 MJ/day to 15.1 MJ/day, and suddenly decreased to 14.5 MJ/day at filling ratio of 70%. This behavior can be attributed to the trade-off between the space available inside the HP and also the heat capacity of the carrying fluid ($m \times C_p$). With the filling ratio increment of the working fluid, the total heat capacity of the working fluid increases, while also the space available for transporting the vapor inside the HP significantly decreases. Hence, there is an optimum value for the filling ratio in which the thermal transport within the HP is maximized, while the heat capacity of the working fluid is plausible as well. Notably, for the tilt angle value, there is another trade-off in between the gravity forces, residence time of the carrying fluid and also the area of the collector exposed to solar insolation. With tilt angle increment, the gravity force facilitates the return of the carrying fluid to the evaporator section while in turn decreases the area exposed to the solar insolation and also decreases the residence time of the carrying fluid in the condenser region. Hence for both parameters, depending on the type of the carrying fluid, the optimum value should be identified.

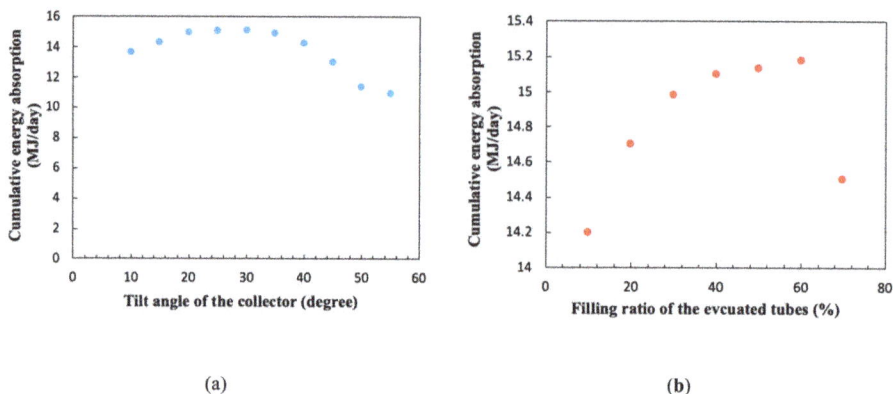

(a)

(b)

**Figure 12.** Dependence of energy absorption by evacuated tube solar collector on tilt angle and filling ratio of the carrying fluid, (**a**) tilt angle, (**b**) filing ratio.

*3.3. Collector's Flow Rate*

Figure 13a–c present the variation of the thermal efficiency of the collector with time for pure acetone and nanofluids at various mass concentration of the carbon nanoparticles and for various flow rates. As can be seen from Figure 13a, with the time spanning, the thermal efficiency increases reaching a peak at 1:00 pm (e.g., 0.7 at flow rate of 3 L/min) and decreases to 0.1 at 8:00 pm interestingly, flow rate of the collector prompted the thermal efficiency of the collector. For example, at 10:00 am, for flow rate of 1 L/min, the thermal efficiency of the collector is 0.31, which reaches 0.5 at flow rate of 3 L/min. For nanofluid, the magnitude of the thermal efficiency increases such that the maximum efficiency reaches from 0.7 to 0.91 for pure acetone and nanofluid at wt. % = 0.1 and at flow rate of 3 L/min. This improvement in the efficiency of the collector can be assigned to the enhancement in the thermal properties of the working fluid inside the evacuated tube and due to the Brownian motion of the carbon nanoparticles inside the heat pipes, which further improves the HTC inside the evaporator especially in the evaporator section.

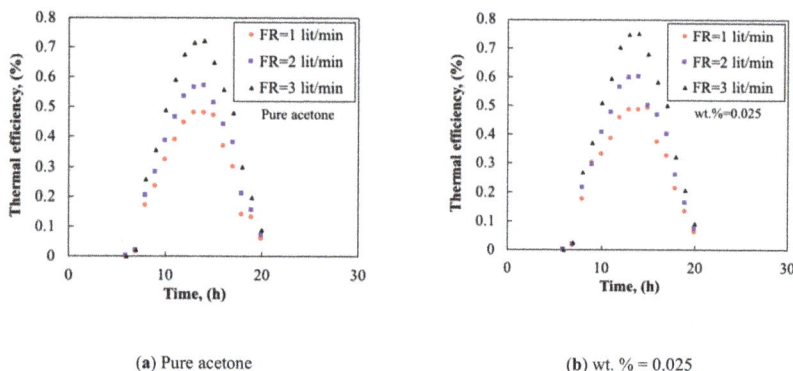

(a) Pure acetone

(b) wt. % = 0.025

**Figure 13.** *Cont.*

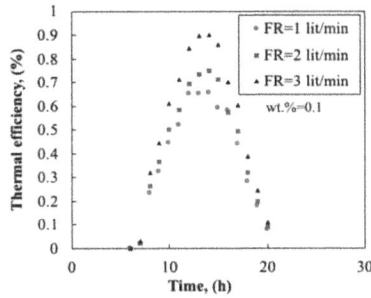

(c) wt. % = 0.1

**Figure 13.** Dependence of thermal efficiency of collector on time for various mass fractions of CNPs in acetone.

*3.4. Temperature Difference*

Figure 14 represents the dependence on time of the temperature difference of the thermal collector for different mass concentrations of the nanofluid and at various flow rates. As can be seen from the figure, an increase in the mass fraction of the carbon nanoparticles can potentially improve the TD value, which in turn improves the TP of the system. For example, the maximum value of TD for the collector working at 1.0 L/min is ~18 °C, which was registered for the nanofluid at wt.% = 0.1. However, the maximum TD was 10 °C and 7.8 °C for flow rates of 2.0 and 3.0 L/min. This can be attributed to the reduction in the residence time inside the collector and inside the header tank, which causes smaller TD values.

Apart from the advantages of the carbon nanoparticles for the collector, the deposition of the carbon nanoparticles on the internal walls of the evaporator might technically be challenging to the continuous operation of the system. Hence, the internal wall of the evaporator of the HP was sent for the scanning electron microscopy to evaluate the potential effect of deposition of CNPs on the internal wall of the evaporator section.

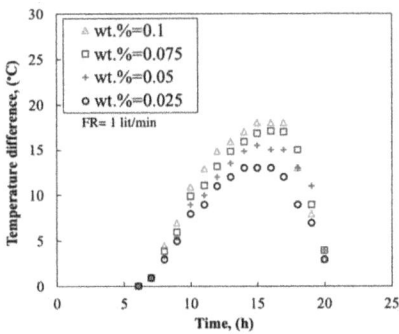

(a) FR = 1 lit/min

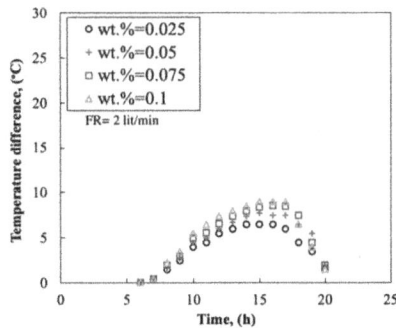

(b) FR = 2 lit/min

**Figure 14.** *Cont.*

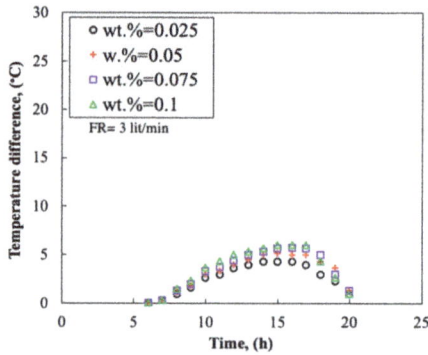

(c) FR = 3 lit/min

**Figure 14.** Variation of the temperature difference with time for various concentration of CNPs.

*3.5. Evaporator's Surface*

Figure 15 presents the SEM images taken from the internal walls of the evaporator of the HP after continuous working with various nanofluids. As can be seen from Figure 15a, a continual layer of deposition was formed on the surface of the evaporator. This layer increased the roughness and irregularities of the surface, which can potentially be some ideal places for the nucleation, and the growth of the bubbles. Interestingly, in Figure 15b, the roughness of the surface together with its morphology changed, which provided different types of the micro-cavities for the nucleation and bubble formation. However, for both surfaces represented in Figure 15a,b, the irregularities and also the non-uniformity of the surface was relatively higher than that of observed for the clean surface presented in Figure 15c. Therefore, despite the stability of the nanofluid, the deposition layer of the nanoparticles forms inside the evaporator section, which is not necessary, a disadvantage. If the main heat transfer mechanism of the evaporator is the nucleate boiling heat transfer, such deposition layers can intensify the TP by generating more bubbles on the surface. However, further research on the mechanism of the deposition of nanoparticles on the surface is required before this finding can be generalized to all types of nanofluids and experimental cases.

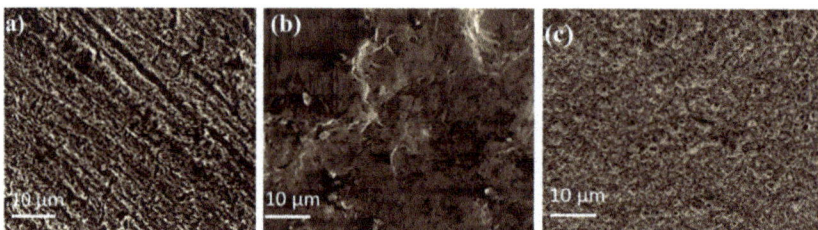

**Figure 15.** SEM images taken from the evaporator after continuous operation with nanofluid, (**a**) wt. % = 0.025, (**b**) wt. % = 0.075, (**c**) clean surface (and also the case of working with acetone had the same roughness shape).

**4. Conclusions**

In the present study, an experimental study was carried out to evaluate the performance of an ETSC working with carbon-acetone and the following remarks were made:

- Carbon nanoparticles were stable up to 10 days within acetone and provided a significant change in the thermo-physical properties of the acetone. The thermal conductivity enhancement of the nanofluid improved the heat transfer mechanism in the HP and enhanced the TP of the collector. The maximum thermal efficiency of 91% was achieved which was well above the average thermal performance of the current collectors working with conventional working fluid (e.g., 72.6% for an acetone solar collector).

- By increasing the mass fraction of the nanofluid, the TP and the TD of the collector increased. The thermal efficiency could reach 0.91 and the maximum TD of 20 °C was obtained at flow rate of 3.0 L/min.
- The deposition of the nanoparticles on the internal wall of the evaporator created an irregular coating on the surface, which was a suitable medium for the nucleation and the formation of the bubbles. Besides the augmentation in the thermal conductivity of acetone, this irregular surface can intensify the bubble formation and their interaction, which resulted in the improvement of the HTC in the evaporator section.

The tilt angle and the filling ratio value were found to influence the cumulative thermal energy of the collector. Hence, the optimum filling ratio of 60% and the optimum tilt angle of 30° were identified, in which the highest TP can be achieved from the collector.

**Author Contributions:** Conceptualization, M.M.S, I.T. and M.R.S.; methodology, I.T. and M.M.S.; validation, M.M.S.; formal analysis, M.M.S. and I.T.; investigation, M.M.S., I.T. and M.A.B.; resources, M.M.S., I.T. and M.A.B.; data curation, M.M.S., M.R.S. and M.A.B.; writing—original draft preparation, M.M.S., M.R.S.; writing—review and editing, M.M.S., I.T., M.R.S. and M.A.B.; visualization, M.M.S, M.R.S.; supervision, I.T. and M.M.S.; project administration, M.R.S.

**Funding:** Mohammad Abdul Baseer would like to thank Deanship of Scientific Research at Majmaah University for supporting this work under the Project Number No. 1440-102.

**Acknowledgments:** The first author appreciates Rain Sanat CO. and Khurshid Bandar CO. For sharing the thermal collector facility.

**Conflicts of Interest:** The authors of the present work declare that there is no conflict of interest regarding the paper.

## References

1. Mahendran, M.; Lee, G.; Sharma, K.; Shahrani, A.; Bakar, R. Performance of evacuated tube solar collector using water-based titanium oxide nanofluid. *J. Mech. Eng. Sci.* **2012**, *3*, 301–310.
2. Sabiha, M.; Saidur, R.; Hassani, S.; Said, Z.; Mekhilef, S. Energy performance of an evacuated tube solar collector using single walled carbon nanotubes nanofluids. *Energy Convers. Manag.* **2015**, *105*, 1377–1388. [CrossRef]
3. Liu, Z.-H.; Hu, R.-L.; Lu, L.; Zhao, F.; Xiao, H.-s. Thermal performance of an open thermosyphon using nanofluid for evacuated tubular high temperature air solar collector. *Energy Convers. Manag.* **2013**, *73*, 135–143. [CrossRef]
4. Tsai, C.; Chien, H.; Ding, P.; Chan, B.; Luh, T.; Chen, P. Effect of structural character of gold nanoparticles in nanofluid on heat pipe thermal performance. *Mater. Lett.* **2004**, *58*, 1461–1465. [CrossRef]
5. Kang, S.-W.; Wei, W.-C.; Tsai, S.-H.; Yang, S.-Y. Experimental investigation of silver nano-fluid on heat pipe thermal performance. *Appl. Therm. Eng.* **2006**, *26*, 2377–2382. [CrossRef]
6. Barber, J.; Brutin, D.; Tadrist, L. A review on boiling heat transfer enhancement with nanofluids. *Nanoscale Res. Lett.* **2011**, *6*, 280–296. [CrossRef] [PubMed]
7. Taylor, R.A.; Phelan, P.E. Pool boiling of nanofluids: Comprehensive review of existing data and limited new data. *Int. J. Heat Mass Transf.* **2009**, *52*, 5339–5347. [CrossRef]
8. Kwark, S.M.; Kumar, R.; Moreno, G.; Yoo, J.; You, S.M. Pool boiling characteristics of low concentration nanofluids. *Int. J. Heat Mass Transf.* **2010**, *53*, 972–981. [CrossRef]
9. Das, S.K.; Choi, S.U.; Patel, H.E. Heat transfer in nanofluids—A review. *Heat Transf. Eng.* **2006**, *27*, 3–19. [CrossRef]
10. Verma, S.K.; Tiwari, A.K. Progress of nanofluid application in solar collectors: A review. *Energy Convers. Manag.* **2015**, *100*, 324–346. [CrossRef]
11. Xuan, Y.; Roetzel, W. Conceptions for heat transfer correlation of nanofluids. *Int. J. Heat Mass Transf.* **2000**, *43*, 3701–3707. [CrossRef]

12. Nakhjavani, M.; Nikkhah, V.; Sarafraz, M.; Shoja, S.; Sarafraz, M. Green synthesis of silver nanoparticles using green tea leaves: Experimental study on the morphological, rheological and antibacterial behaviour. *Heat Mass Transf.* **2017**, *53*, 3201–3209. [CrossRef]
13. Nikkhah, V.; Sarafraz, M.; Hormozi, F. Application of spherical copper oxide (ii) water nano-fluid as a potential coolant in a boiling annular heat exchanger. *Chem. Biochem. Eng. Q.* **2015**, *29*, 405–415. [CrossRef]
14. Salari, E.; Peyghambarzadeh, S.; Sarafraz, M.; Hormozi, F.; Nikkhah, V. Thermal behavior of aqueous iron oxide nano-fluid as a coolant on a flat disc heater under the pool boiling condition. *Heat Mass Transf.* **2017**, *53*, 265–275. [CrossRef]
15. Sarafraz, M. Experimental investigation on pool boiling heat transfer to formic acid, propanol and 2-butanol pure liquids under the atmospheric pressure. *J. Appl. Fluid Mech.* **2013**, *6*, 73–79.
16. Sarafraz, M.; Hormozi, F.; Peyghambarzadeh, S.; Vaeli, N. Upward flow boiling to di-water and cuo nanofluids inside the concentric annuli. *J. Appl. Fluid Mech.* **2015**, *8*, 651–659.
17. Sarafraz, M.; Nikkhah, V.; Madani, S.; Jafarian, M.; Hormozi, F. Low-frequency vibration for fouling mitigation and intensification of thermal performance of a plate heat exchanger working with cuo/water nanofluid. *Appl. Therm. Eng.* **2017**, *121*, 388–399. [CrossRef]
18. Sarafraz, M.; Nikkhah, V.; Nakhjavani, M.; Arya, A. Thermal performance of a heat sink microchannel working with biologically produced silver-water nanofluid: Experimental assessment. *Exp. Therm. Fluid Sci.* **2018**, *91*, 509–519. [CrossRef]
19. Sarafraz, M.; Peyghambarzadeh, S.; Alavi Fazel, S.; Vaeli, N. Nucleate pool boiling heat transfer of binary nano mixtures under atmospheric pressure around a smooth horizontal cylinder. *Period. Polytech. Chem. Eng.* **2013**, *57*, 71–77. [CrossRef]
20. Sarafraz, M.; Peyghambarzadeh, S. Nucleate pool boiling heat transfer to al2o3-water and tio2-water nanofluids on horizontal smooth tubes with dissimilar homogeneous materials. *Chem. Biochem. Eng. Q.* **2012**, *26*, 199–206.
21. Sarafraz, M.; Arjomandi, M. Demonstration of plausible application of gallium nano-suspension in microchannel solar thermal receiver: Experimental assessment of thermo-hydraulic performance of microchannel. *Int. Commun. Heat Mass Transf.* **2018**, *94*, 39–46. [CrossRef]
22. Sarafraz, M.; Arjomandi, M. Thermal performance analysis of a microchannel heat sink cooling with copper oxide-indium (cuo/in) nano-suspensions at high-temperatures. *Appl. Therm. Eng.* **2018**, *137*, 700–709. [CrossRef]
23. Sarafraz, M.; Arya, H.; Arjomandi, M. Thermal and hydraulic analysis of a rectangular microchannel with gallium-copper oxide nano-suspension. *J. Mol. Liq.* **2018**, *263*, 382–389. [CrossRef]
24. Sarafraz, M.; Arya, H.; Saeedi, M.; Ahmadi, D. Flow boiling heat transfer to mgo-therminol 66 heat transfer fluid: Experimental assessment and correlation development. *Appl. Therm. Eng.* **2018**, *138*, 552–562. [CrossRef]
25. Sarafraz, M.; Hart, J.; Shrestha, E.; Arya, H.; Arjomandi, M. Experimental thermal energy assessment of a liquid metal eutectic in a microchannel heat exchanger equipped with a (10 hz/50 hz) resonator. *Appl. Therm. Eng.* **2019**, *148*, 578–590. [CrossRef]
26. Dehaj, M.S.; Mohiabadi, M.Z. Experimental investigation of heat pipe solar collector using mgo nanofluids. *Sol. Energy Mater. Sol. Cells* **2019**, *191*, 91–99. [CrossRef]
27. Ziyadanogullari, N.B.; Yucel, H.; Yildiz, C. Thermal performance enhancement of flat-plate solar collectors by means of three different nanofluids. *Therm. Sci. Eng. Prog.* **2018**, *8*, 55–65. [CrossRef]
28. Sharafeldin, M.; Gróf, G. Efficiency of evacuated tube solar collector using wo3/water nanofluid. *Renew. Energy* **2019**, *134*, 453–460. [CrossRef]
29. Sharafeldin, M.; Gróf, G. Evacuated tube solar collector performance using ceo 2/water nanofluid. *J. Clean. Prod.* **2018**, *185*, 347–356. [CrossRef]
30. Ozsoy, A.; Corumlu, V. Thermal performance of a thermosyphon heat pipe evacuated tube solar collector using silver-water nanofluid for commercial applications. *Renew. Energy* **2018**, *122*, 26–34. [CrossRef]
31. Nazari, M.A.; Ghasempour, R.; Ahmadi, M.H.; Heydarian, G.; Shafii, M.B. Experimental investigation of graphene oxide nanofluid on heat transfer enhancement of pulsating heat pipe. *Int. Commun. Heat Mass Transf.* **2018**, *91*, 90–94. [CrossRef]
32. Arya, A.; Sarafraz, M.; Shahmiri, S.; Madani, S.; Nikkhah, V.; Nakhjavani, S. Thermal performance analysis of a flat heat pipe working with carbon nanotube-water nanofluid for cooling of a high heat flux heater. *Heat Mass Transf.* **2018**, *54*, 985–997. [CrossRef]

33. Vijayakumar, M.; Navaneethakrishnan, P.; Kumaresan, G. Thermal characteristics studies on sintered wick heat pipe using cuo and al2o3 nanofluids. *Exp. Therm. Fluid Sci.* **2016**, *79*, 25–35. [CrossRef]
34. Brunauer, S.; Emmett, P.H.; Teller, E. Adsorption of gases in multimolecular layers. *J. Am. Chem. Soc.* **1938**, *60*, 309–319. [CrossRef]
35. Gor, G.Y.; Thommes, M.; Cychosz, K.A.; Neimark, A.V. Quenched solid density functional theory method for characterization of mesoporous carbons by nitrogen adsorption. *Carbon* **2012**, *50*, 1583–1590. [CrossRef]
36. Sarafraz, M.; Hormozi, F. Experimental study on the thermal performance and efficiency of a copper made thermosyphon heat pipe charged with alumina–glycol based nanofluids. *Powder Technol.* **2014**, *266*, 378–387. [CrossRef]
37. Sarafraz, M.; Hormozi, F. Heat transfer, pressure drop and fouling studies of multi-walled carbon nanotube nano-fluids inside a plate heat exchanger. *Exp. Therm. Fluid Sci.* **2016**, *72*, 1–11. [CrossRef]
38. Sarafraz, M.; Hormozi, F.; Peyghambarzadeh, S. Role of nanofluid fouling on thermal performance of a thermosyphon: Are nanofluids reliable working fluid? *Appl. Therm. Eng.* **2015**, *82*, 212–224. [CrossRef]
39. Sarafraz, M.; Hormozi, F.; Peyghambarzadeh, S. Thermal performance and efficiency of a thermosyphon heat pipe working with a biologically ecofriendly nanofluid. *Int. Commun. Heat Mass Transf.* **2014**, *57*, 297–303. [CrossRef]
40. Sarafraz, M.; Safaei, M.R. Diurnal thermal evaluation of an evacuated tube solar collector (ETSC) charged with graphene nanoplatelets-methanol nano-suspension. *Renew. Energy* **2019**. In press, accepted manuscript. [CrossRef]
41. Moffat, R.J. Describing the uncertainties in experimental results. *Exp. Therm. Fluid Sci.* **1988**, *1*, 3–17. [CrossRef]

*applied sciences*

MDPI

*Article*

# Impedance Measurement and Detection Frequency Bandwidth, a Valid Island Detection Proposal for Voltage Controlled Inverters

**Marc Llonch-Masachs, Daniel Heredero-Peris, Cristian Chillón-Antón, Daniel Montesinos-Miracle and Roberto Villafáfila-Robles**

Centre d'Innovació Tecnològica en Convertidors Estàtics i Accionaments (CITCEA-UPC), Departament d'Enginyeria Elèctrica, Universitat Politècnica de Catalunya, ETS d'Enginyeria Industrial de Barcelona, Av. Diagonal, 647, Pl. 2. 08028 Barcelona, Spain; daniel.heredero@citcea.upc.edu (D.H.-P.); cristian.chillon@citcea.upc.edu (C.C.-A.); montesinos@citcea.upc.edu (D.M.-M.); roberto.villafafila@citcea.upc.edu (R.V.-R.)
* Correspondence: marc.llonch@citcea.upc.edu; Tel.: +34-934-016-855

Received: 27 February 2019; Accepted: 13 March 2019; Published: 18 March 2019

**Abstract:** Anti-islanding detection methods have been part of a secure operation for distributed energy resource inverters, avoiding the creation of non-intentional energization when the mains are lost. These detection mechanisms were conceived historically for current-controlled inverters. New control possibilities have broken ground, and current- or voltage-controlled inverters are a reality; however, special attention must be paid to detection strategies when applied to the latter ones. This paper addresses two topics: it exposes the lack of effectiveness of those detection algorithms based on the voltage/frequency displacement concept under voltage-controlled inverters and evaluates the applicability limits of the others based on the impedance measurement (IM). The IM is presented as a valid mechanism to achieve the islanding detection, but the exploration of its limits drives the concept of detection frequency bandwidth (DFBW), introduced in this paper. The DFBW is suggested as a practical approach to select the proper injection frequency to measure. Therefore, an improved strategy based on the IM and DFBW is proposed to allow achieving the detection towards (non-)resonant loads considering low computational burden. The results were experimentally validated in a 90-kVA four-wire voltage-controlled inverter, offering detection times of less than 100 ms in any case.

**Keywords:** anti-islanding; voltage source converters; impedance measurement; microgrids

---

## 1. Introduction

Utility deregulation has sped up the possibilities for distributed energy resources (DER). The high penetration level of DERs is not only a well-consolidated reality today, but is also experiencing a steady rise thanks to a global environmental awareness of clean energies. A clear picture of this fact is their percentage share of the energy mix [1]. Although most DER inverters were conceived of in the past to behave as CC-VSI (current-controlled voltage source inverters), this situation is evolving. New control techniques, the use of information technologies, and the progress of power electronics have boosted alternative options for either grid-connected or grid-disconnected operation. Other behaviors like voltage-controlled (VC)-VSI inverters enhance grid flexibility services such as power sharing or non-zero voltage crossing between operation modes [2–4]. Currently, it is possible to distinguish between three main types of grid integration inverters [5]:

-   Grid supply inverters (GSI) are unidirectional CC-VSI when grid-connected, and their aim is to deliver the maximum power to the mains by means of maximum power point tracking

(MPPT) algorithms [6]. They are the most common inverters installed in grid-connected systems, but always require a voltage master source for the energy exchange. As the situation of always delivering its maximum power is excessively rigid, some regulations like VDE 4105 [7] (Verband der Elektrotechnik, Elektronik und Informationstechnik) permit the grid-operator to manage the exchanged power by modifying the present AC system frequency.

- Grid constitution inverters (GCI) are ideal VC-VSIs and act as the voltage reference source for GSIs in grid-disconnected operation. In grid-connected operation, it can be used as a passive back-up system, with no exchange of power, prepared to act in case of a black-out. In other words, during grid-connected operation, it acts as a mirror of the mains. However, this means that most of the time, a GCI is just wasting its internal losses.
- Grid support and constitution inverters (GSCI) [8–10] are non-ideal VC-VSIs based on the AC droop control. They can respond in both operation modes (grid-connected and grid-disconnected) as voltage source, minimizing the impact on the transference in between modes. A GSCI manages the exchange power by establishing a delta of the amplitude and phase with respect to the mains in grid-connected mode. In grid-disconnected mode, they continue operating as VC-VSI, self-generating the amplitude and phase with the so-called secondary control.

Any of the previously listed inverter types (GSI, GCI, or GSCI) can still be operating in grid-connected mode when an unintentional grid disconnection occurs, a situation called islanding in the literature. Islanding situations put the involved installation (inverter included) in a vulnerable position in terms of safety because of the possibility of the inverter holding the energization of a grid section, so much so that anti-islanding (AI) regulations can be found such as VDE 4105, IEEE 1547, or IEC 61727 [7,11,12], facing how to respond to this situation. These regulations define the threshold times to detect the islanding occurrence in order to stop feeding the fault and the times for considering a stable recovery of the mains to proceed with a following reconnection. The corresponding detection times are summarized in Table 1.

**Table 1.** Summary of islanding threshold detection times.

| Standard | Voltage Threshold (%) | Time (s) | Frequency Threshold (Hz) | Time (s) |
|---|---|---|---|---|
| VDE [a] 4105 | $U < 80$ | 0.20 | $f < 47.5$ | 0.20 |
| | $80 \leq U \leq 115$ | 5.00 | $47.5 \leq f \leq 51.5$ | 5.00 |
| | $U > 115$ | 0.20 | $f > 51.5$ | 0.2 |
| | | | $f^1 < 59.3$ | 0.16 |
| | $U < 50$ | 0.16 | $59.3 \leq f^1 \leq 60.5$ | 2.00 |
| IEEE [b] 1547 | $50 \leq U < 88$ | 2.00 | $f^1 > 60.5$ | 0.16 |
| | $88 \leq U \leq 110$ | 2.00 | $f^2 < 57.0$ | 0.16 |
| | $110 < U < 120$ | 1.00 | $57.0 \leq f^2 < 59.8$ | 0.16–300 [3] |
| | $U > 120$ | 0.16 | $59.8 \leq f^2 \leq 60.5$ | 2.00 |
| | | | $f^2 > 60.5$ | 0.16 |
| | $U < 50$ | 0.10 | | |
| IEC [c] 61727 | $50 \leq U < 85$ | 2.00 | $f < 49.0$ | 0.20 |
| | $85 \leq U \leq 110$ | 2.00 | $49.0 \leq f \leq 51.0$ | 2.00 |
| | $110 < U < 135$ | 2.00 | $f > 51.0$ | 0.20 |
| | $U > 135$ | 0.05 | | |

[1] DER size $\leq$ 30 kW; [2] DER size $>$ 30 kW; [3] adjustable. [a] Verband der Elektrotechnik, Elektronik und Informationstechnik. [b] Institute of Electrical and Electronics Engineers. [c] Intenrnational Electrotechnical Comission.

Putting AI detection methods (AIDM) into historical context, they appeared when the renewable integration did not consider islanded mode feeding options. Therefore, it is no wonder that AI standards were targeted for CC-VSIs, being the control type most widely used in the case of conventional photovoltaic farms or wind farms. However, new grid integration roles have emerged, demanding more VC-VSIs. In the next few years, electrical distribution grids are expected to be smarter due to the current changing energy scenario. New concepts related to efficiency, reliability, robustness, management, and business are finding their way, such as microgrids [13,14], smart-grids [15], energy hubs [16], power routers [17], or virtual power plants [18]. In the mentioned new framework, GSIs will start to coexist in the grid with other devices such as GCIs and GSCIs. Hence, VC-VSIs will be required, and their implications for AI detection need to be analyzed and incorporated.

The present paper aims to face the challenge of using AIDM with VC-VSI, mainly focusing on GSCI, because of their future expected participation in DER integration. To the best knowledge of the authors, this has resulted in it being difficult to find works dealing with the interactions between AIDM and VC-VSIs. In fact, recent literature shows the same trend as in the past [19–24]. These authors devoted efforts to enhance previously-existing AIDM, analyzing the effect under different grid faults or considering the criticality of the loads, but regularly assuming the inverter as a CC-VSI. The only exceptions found are not always focused on power electronics. Some authors as in [25,26] mentioned AIDM for diesel generators, but this AIDM is based on remote strategies, in which the AI detector is not part of the generator itself. The present authors have published some preliminary papers in line with the challenge of AIDM for VC-VSI in [27,28].

This paper addresses two main contributions. Firstly, we explore the lack of effectiveness of islanding detection methods based on voltage and frequency displacements applied to VC-VSI, paying special attention to GSCI. In this sense, the conventional definition of resonant load present in the literature should be reconsidered. In this work, it is redefined from a zero power flow point of view, extending its applicability to any kind of inverter (whether it be VC or CC).

Secondly, and according to the first part of the paper, IM strategies are chosen to overcome island detection in any inverter type, but their limitations are analyzed. One of the main contributions of the present article resides in the use of a new concept called DFBW (detection frequency bandwidth), which facilitates judging the performance of islanding detection depending on the local load type and the strength of the mains. For this purpose, an AIDM for VC-VSI based on the IM variation concept is applied. This method considers selecting appropriately a perturbation frequency according to the DFBW. The used AI method is based on [29], but it has been modified to lessen the computational burden, being another of the key contributions. The final AI method proposed is validated in an experimental setup considering a three-phase four-wire 90-kVA GSCI, obtaining detection times of 100 ms in compliance with VDE 4105, IEEE 1547, or IEC 61727 [7,11,12].

Thus, the paper is structured as follows. Section 2 introduces both VC-VSI and CC-VSI and extends the concept of (non-)resonant load to (non)zero power flow for generalization. Then, Section 3 summarizes the main AIDMs present in the literature. In Section 4, the main AIDM limitations when applied to VC-VSI are explored. This leads to Section 5, in which the proposed solution based on IM is presented. This proposal is evaluated experimentally in Section 6. Finally, Section 7 points out the derived conclusions.

## 2. Control Fundamentals

Firstly, before presenting the interactions between AI mechanisms and VSIs, it is significant to address its possible voltage and current source behaviors, referred to as VC-VSI and CC-VSI, respectively. The strict concept of resonant load detailed in the AI regulations is extended to (non)zero power flow ((N)ZPF) for generalization.

### 2.1. Control Type for VSI

A VC-VSI requires, at least, disposing of an LC-type (inductive-capacitive) filter at its output; see Figure 1a. The AC capacitor voltage, $U_C$, is managed by a voltage control loop. Its control action regulates the reference of the inner current loop, $I_{L_1}^*$, ensuring the converter current during operation. Usually, a second inductance is also considered, $L_2$, which represents the aggregation of the leakage inductance of a transformer and the line impedance between the capacitor and the point of common coupling (PCC). This impedance makes possible a third master control loop corresponding to the AC droop control in GSCI. This allows handling the power exchange with the mains. Its control action determines the amplitude and phase for the voltage reference $U_C^*$. Accordingly, the aim of using an AC droop control strategy is to allow the operation of several VC-VSI in parallel, mainly without the use of communications [8–10]. AC droop control is usually considered as a non-ideal voltage source because its apparent power regulation is achieved by managing the difference of the frequency and voltage amplitude between $U_C$ and $U_{PCC}$.

On the other hand, a CC-VSI only strictly requires an inductive output filter. Apparent power can also be controlled, but it is directly regulated according to a translation from the apparent power reference to amplitude and phase in terms of current, as shown in Figure 1b. The active or reactive power amount can be simply translated to current by the ratio of the reference over the instant voltage or the instantaneous voltage lagging a grid quarter period, respectively.

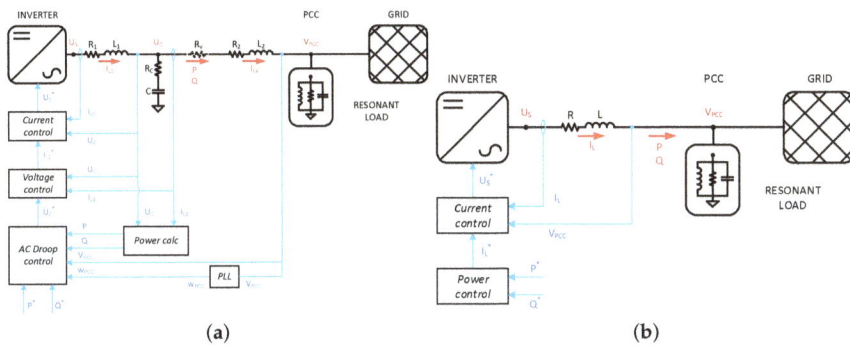

**Figure 1.** Control topology in grid-connected mode with a resonant local load. (**a**) Voltage-controlled voltage source inverter (VC-VSI) topology. (**b**) Current-controlled voltage source inverter (CC-VSI) topology. PCC, point of common coupling; PLL, phase locked loop.

### 2.2. Zero Power Flow Concept

Figure 1 shows an inverter connected to the grid with the resonant load type proposed in the mentioned AI standards [7,11,12]. They suggest that the concept of resonant load take place when the local consumption is nearly equal to the active power given by the inverter in terms of current, considering no reactive power, but with a resonance between the reactive components equal to the frequency of the grid. On the one hand, the active part is modeled by a lumped resistance. On the other hand, an inductance is parallelized with a capacitance to achieve the resonance frequency.

It should be highlighted that when the load is under resonance, there is no reactive consumption. Note that the resonance concept is linked to a voltage driven from the current delivered by the inverter. This conventional resonant concept is limited and less significant for VC-VSI where the voltage at the PCC is explicitly controlled by the inverter. In other words, this is due to the voltage and frequency being maintained by the control law of the inverter. Thus, when using VC-VSIs, the voltage loop stunts the mains loss detection. Note that these situations are possible even though the PCC power flow is nonzero. When the mains are lost, the voltage of the PCC, $U_{PCC}$, is not released and becomes conditioned by the inverter. When the grid is off, the grid inertia is free, and $U_{PCC}$ becomes indirectly controlled by the inverter. As can be seen in Figure 1a, the AC capacitor voltage $U_C$ and the $U_{PCC}$ voltage are connected only through the output coupling impedance constituted by the inductance $L_2$ and its equivalent series resistance $R_2$, which are usually small in value.

According to the exposed arguments, it is appropriate to reconsider the resonant load concept. The resonant load purpose is to be the worst scenario for the AI methods based on voltage and frequency displacements (AI-PM (Passive Methods)and AI-PF (Positive Feedback)), refer to Section 3 for better understanding. Thus, the worst scenario occurs assuming the power flow through the PCC (active and passive) to be about zero [28,30]. Accordingly, the conception of zero (ZPF) or non-zero power flow (NZPF) exposed points out the resonant condition.

### 3. Main Anti-Islanding Detection Methods

Anti-islanding detection methods (AIDM) have been evolving during the last few decades. A summary picture of the main AIDM can be seen in Figure 2, in which two categories can be defined: local or remote, according to if the AIDM is or is not part of the inverter itself. Then, three sub-categories can be considered; passive methods (PM), active methods (AM), and methods based on communications (C).

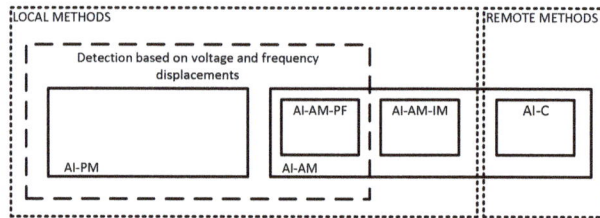

**Figure 2.** Anti-islanding detection methods' classification. PM, passive methods; AM, active methods; C, communications; IM, impedance measurement; PF, positive feedback.

The first adopted solution, called passive methods (PM), is based on the observation of the point of common coupling (PCC) (refer to Figure 1) to detect the loss of mains by means of monitoring the voltage or the frequency variations [31–33]. Passive detection methods (PM) operate by monitoring the voltage amplitude and the frequency of the PCC defining a non-detection zone (NDZ). The NDZ is essentially a window defined by the voltage amplitude and frequency thresholds where the inverter is not able to detect any island event. When the mains are off, due to any external occurrence (intentional or non-intentional), the voltage or the frequency could be pushed enough to exceed NDZ limits, as is described in Figure 3.

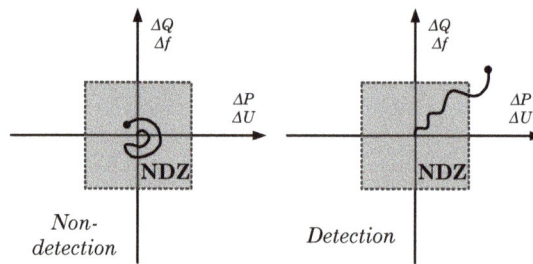

**Figure 3.** Non-detection zone (NDZ).

However, if a local load, usually called resonant load, is consuming all the power (active and reactive) that the inverter is delivering when the utility goes off, neither the voltage nor the frequency will change significantly. This fact hinders the island detection. Thus, active methods (AM) appeared to overcome this challenge. Their aim is to observe, as well as to provoke a perturbation, boosting the variation of some electric variable, consequently enhancing that observation. According to their operating principle, they can be classified into three main groups:

- Positive feedback (PF): These are active methods injecting some perturbation in order to generate a voltage or frequency drift to make evident the islanding transition. Several examples can be found in the literature: active frequency drift (AFD) [31,32], slip mode frequency shift (SMS) [31–34], Sandia voltage shift (SVS), and Sandia frequency shift (SFS) [31,32,35].

- Impedance measurement (IM): An impedance measurement mechanism detects the utility disconnection thanks to the impedance change of the PCC. Some authors started using external switching impedances devices [36,37], but the increase of the computational capabilities of the inverters microcontrollers makes them able to inject harmonics under different approaches: simple harmonic injection [31,38], double harmonic injection [31,38], and also using inter-harmonics [39–41]. Finally, as indirect harmonic injection, PLL (phase locked loop) can also be found [29,31], as well as system identification-based methods [42].

- Based on communications [25,26]: Some signal generator based on different carrying means aims to provide the state signals of the mains. Some examples of such possibilities are radio, power line carrier communication, power signals, or wireless alternatives. They usually need a remotely located transmitter and receiver, and this responds to why it is considered a remote AIDM. However, it is possible to find some examples based on a power device itself under a local conceptualization [25].

## 4. Limitations of AIDMs When Applied with the Presence of GSCI

This section aims to expose the limited capacities or constraints from those AI-PM and AI-AM locally embedded in a GSCI. AI-C are out of the scope of the present paper because they usually imply remote requirements. Thus, this section leads to answering why the AI-IM method results in a proper solution and contributes to analyzing the useful frequencies to apply to it. For this purpose, the frequency detection bandwidth (DFBW) will be introduced and evaluated according to the impedance PCC transfer function.

### 4.1. Passive Method Type

A droop control strategy is usually used in inverters (GSCI case) due to its ease of parallelization [4,5,9,10]. The droop control manages the active and reactive power, modifying the module and phase of the AC capacitor voltage $U_C$ [9,10] with respect to the PCC voltage $U_{PCC}$. Nevertheless, for its proper operation, the droop control is based on grid-connected operation and needs the grid to hold the voltage of the PCC (module and phase). Due to this, the droop control requires great effort to change the injected power, active or reactive, under the absence of the mains. In other words, when the voltage loop's control action attempts to correct $U_C$, so that it varies the power flow exchange with $U_{PCC}$, the $U_{PCC}$ follows $U_C$, which results in displacement. This phenomenon is qualitatively represented in Figure 4 by a mechanical analogy. In grid-connected mode, the grid is the point of support, i.e., acts as a wall/ground surface. Thus, the phase and amplitude between $U_C$ and $U_{PCC}$ are the "forces" connected to the point of support by rotational and linear springs, respectively. These springs are related to the impedance between $U_C$ and $U_{PCC}$. The more "elongation" there is or the more "angle rotated" it is, the more "force" results (active or reactive power managed by means of droop phase or amplitude action). In the case of grid absence, the point of support is lost, and any effort in gaining elongation or angle by $U_C$ will consequently drag $U_{PCC}$.

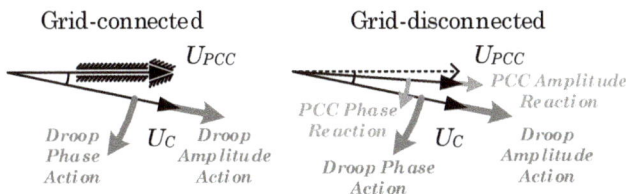

**Figure 4.** Vectorial representation of droop control under a grid disconnection.

Although droop's control actions are affected by the sudden power flow interruption of grid-connected to grid-disconnected transitions, the effects result in being excessively smooth. This is due to the implicit slow dynamic behavior when emulating synchronous generator inertias [5]. Thus, due to a reduced time response, as well as limited droop control action's swing (in terms of voltage amplitude and phase), PM are not adequate because several standards require a detection in less than 160 ms, as will be detailed in Section 4.2.4. Thus, AI-PM shows an implicit lack of effectiveness in the conventional concept of resonant load and NZPF and ZPF.

### 4.2. Active Method-Positive Feedback Type

In order to boost the potential of island detection, AI-AM are widely used in contrast to AI-PM. In the following lines, why AM-PF characterized by voltage and frequency shifts are not adequate when ZPF is assumed in VC-VSI is detailed. Thus, the following lines put the focus on the lack of effectiveness of the main AM-PF options; slip mode shift (SMS), active frequency drift (AFD), and Sandia voltage or Sandia frequency shift (SVS or SFS).

#### 4.2.1. Operating Principle

Any AI-AM establishes different mechanisms to perturb, at least, one of the controlled arguments of an electric magnitude. The direct consequence is an intrinsic distortion of the waveform quality. Considering that the inner controlled magnitude is the inverter's output current, $I_{L1}$ (see Figure 1), it can be time-varying, described as:

$$i_{L1}(t) = I \cdot sin(\omega t + \phi) \tag{1}$$

Thus, according to the perturbation strategy, it is possible to change the peak amplitude of the controlled current $I$, the frequency $\omega$, or the phase $\phi$. Some AM-AI methods are called PF methods because their objective is try to make unstable, by some means, the controlled magnitude. The use of the current is the option addressed more often in the literature for AM-PF [31–35].

When the inverter goes from grid-connected to grid-disconnected, the behavior of the voltage and frequency at the PCC is directly related to the impedance and the resonant load power consumption. In terms of the quality factor, $q$, and the resonant angular frequency, $\omega_r$, obtained as:

$$q = R_{RL} \cdot \sqrt{\frac{C_{RL}}{L_{RL}}}$$
$$\omega_r = \frac{1}{\sqrt{L_{RL} \cdot C_{RL}}} \tag{2}$$

the resonant load impedance and its phase can be expressed by:

$$\vec{Z}_{RL} = R_{RL} \cdot \frac{1}{1 + j \cdot q \left( \frac{\omega}{\omega_r} - \frac{\omega_r}{\omega} \right)}$$
$$\theta_{RL} = \arctan \left( -q \cdot \left( \frac{\omega}{\omega_r} - \frac{\omega_r}{\omega} \right) \right) \tag{3}$$

where $R_{RL}$, $L_{RL}$, and $C_{RL}$ are the resistance, the inductance, and the capacitor that characterize the classical definition of a resonant load, respectively. Otherwise, the active, $P_{RL}$, and reactive, $Q_{RL}$, power consumption of this kind of resonant load can be expressed as:

$$P_{RL} = U_{PCC}^2 \cdot \frac{1}{R_{RL}} \tag{4}$$

$$Q_{RL} = U_{PCC}^2 \cdot \left( \frac{1}{\omega \cdot L_{RL}} - \omega \cdot C_{RL} \right) \tag{5}$$

which can be represented in terms of $q$ and $\omega_r$, by:

$$Q_{RL} = U_{PCC}^2 \cdot \frac{q}{R_{RL}} \cdot \left( \frac{\omega_r}{\omega} - \frac{\omega}{\omega_r} \right) \tag{6}$$

Figure 5a shows the active power versus voltage and will be used as the basis plot for Figure 5b–d in which the quality factor $q$ is considered. It can be seen in Figure 5b–d, deduced from Equations (3) and (6), that the impedance (module and phase) and the reactive power slopes close to the resonant frequency are highly sensible with the quality factor $q$. Thus, it is a significant effect that the quality factor $q$ affects AI strategies based on voltage and frequency displacements. The higher the quality factor $q$, the bigger will be the reactive power mismatch needed to move the frequency.

Notice that when grid-disconnected, AM-PF are acting on the local load characteristics to push the PCC voltage or frequency from the NDZ.

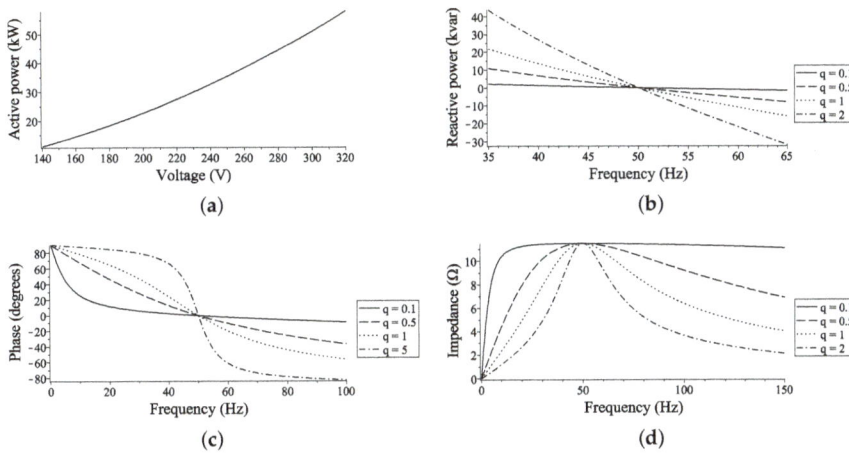

**Figure 5.** Resonant load characteristics. (**a**) Active power versus voltage of a 50-Hz resonant load. (**b**) Reactive power versus frequency of a 50-Hz resonant load. (**c**) Phase versus frequency of a 50-Hz resonant load. (**d**) Impedance versus frequency of a 50-Hz resonant load.

### 4.2.2. Slip Mode Frequency Shift

SMS is an AM-PF that tries to destabilize the inverter by changing the frequency of the delivered current [31–34]. It is based on the phase characteristic of the conventional resonant load concept. Thus,

$$\theta_{RL} = \arctan \left[ -q \cdot \left( \frac{f}{f_r} - \frac{f_r}{f} \right) \right] \tag{7}$$

where $\theta_{RL}$ is the present phase in grid-disconnected mode absorbed by the resonant load, $f$ is the frequency at the PCC, and $f_r$ and $q$ are the resonant frequency and quality factor of the local resonant load, respectively. In order to make the frequency unstable, a perturbation is added on the current injected phase,

$$\theta_{SMS} = \theta_m \cdot \sin \left( \frac{\pi}{2} \cdot \frac{f - f_r}{f_m - f_r} \right) \tag{8}$$

where $\theta_{SMS}$ is strictly the phase of the set-point current and $\theta_m$ and $f_m$ are the SMS parameters.

From Equations (7) and (8), it can be deduced that when grid-connected, the utility holds the frequency about its fundamental value [31]. Even so, when grid-disconnected, this is an unstable equilibrium point, and the frequency is forced to move out from the NDZ. However, SMS has to be injected by means of current directly to the resonant load, taking into account that the outermost controlled variable is $U_C$. This AM results in being meaningless for GSCI.

### 4.2.3. Active Frequency Drift

AFD is a PF method that shares the same aim as SMS [31,32]. The inverter current is distorted with a small zero current segment. Then, the current trends to change the frequency held by the grid. However when grid-disconnected, the frequency is drifted away from the NDZ. Once again, as like SMS, AFD has to be injected by means of current directly to the resonant load, and it is useless for VC-VSI.

### 4.2.4. Sandia Voltage Shift and Sandia Frequency Shift

SVS and SFS are two important AM based on PF strategies [31,32,35]. As a difference from SMS and AFD, the used perturbations in these cases are not directly based on current. Derived from Equations (4) and (5), these methods aim to destabilize the voltage or the frequency by modifying the active or reactive power loops. For the following explanation, consider Figure 1a.

In the SVS case, the variation between the PCC voltage, $U_{PCC}$, and the rated voltage, $U_N$, is accumulated to the active power reference amplified by a gain factor, $k_U$, as is shown in Figure 6a. $G_{PU}(s)$ represents the transfer function, including the inverter, between the active power error and $U_{PCC}$ amplitude. If the grid is connected, the voltage variation is limited and small, offering a null average value ($\Delta P^* \simeq 0$). Then, it affects only slightly the power reference, $P^*$. If the grid is disconnected, any minor voltage discrepancy boosts (in positive or negative direction) the active power reference, which in the next algorithm iteration, will speed up $U_{PCC}$ even more.

The same idea of SVS can be exported to SFS in terms of frequency considering the negative slope of the relation between the reactive power and the frequency (refer to Figure 5b), as can be seen in Figure 6b. In this case, the amplifier factor is $k_\omega$, and $G_{Q\omega}(s)$ represents the transfer function, including the inverter, between the reactive power error and $\omega_{PCC}$.

**Figure 6.** AI-PF methods. (**a**) Sandia voltage shift (SVS). (**b**) Sandia frequency shift (SFS).

If the SFS option is simulated under a GSI or a GSCI context, it can be seen how this fact is translated to extremely high positive feedback amplifier gains. The inverter and control parameters used to evaluate this discrepancy are summarized in Table 2. In order to simulate CC-VSI, the voltage loop and the droop control have been removed.

Figure 7a (CC-VSI) and Figure 7b (VC-VSI) represent the detection time when the inverter delivers 30 kW. The grid to which it is connected is 230 V and 50 Hz. The resonant load is 30 kW and the quality factor $q = 2$ is assumed ($R_{RL} = 1.76\ \Omega$; $L_{RL} = 2.8$ mH; $C_{RL} = 3.6$ mF). As previously mentioned, high amplifier gains produce non-convenient disturbances. Figure 7a,b shows the islanding detection times versus the mains' frequency at different $k_\omega$ factors. Figure 7a shows slow detection times as closest to the rated frequency (50 Hz in the example), which is the frequency at the PCC. This is due to the small initial error of the positive feedback loop. Even considering this behavior, the detection times results in being fast enough to achieve the proposed time in the standard. In contrast, in Figure 7b, it can be noted that the response for a VC-VSI implies higher amplifying gains to achieve acceptable detection times without guaranteeing the aforementioned threshold times. Analogous results can be obtained using SVS instead of SFS.

**Table 2.** Hardware and control parameters.

|  |  |  |  |
|---|---|---|---|
| LCLfilter | Output phase inductance $L_1$ | 250 | µH |
|  | Transformer leakage inductance $L_2$ | 70 | µH |
|  | AC capacitor $C$ | 350 | µF |
|  | Switching frequency | 8 | kHz |
| Droop loop | Proportional constant for $P$, $Q$ | 0.00001 |  |
|  | Integral constant for $Q$ | 0.003 |  |
| Voltage loop | PRproportional constant | 0.07 |  |
|  | PR integral constant | 0.07 |  |
| Current loop | PR proportional constant | 0.75 |  |
|  | PR integral constant | 3.93 |  |
| PLL loop | Settling time | 5 | ms |

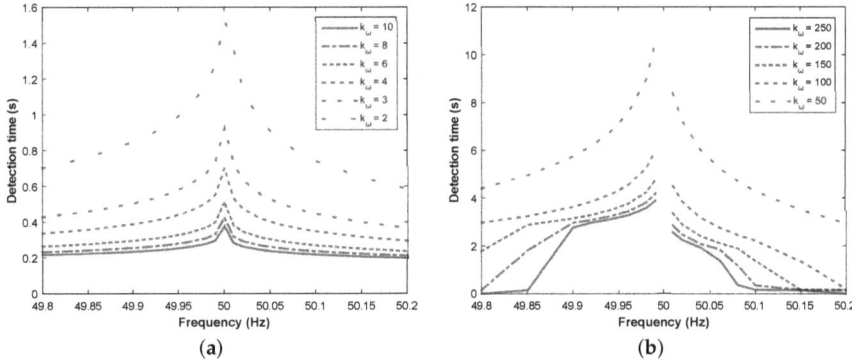

**Figure 7.** SFS detection times applied to CC-VSI and VC-VSI. (**a**) Operating with a grid supply inverter (GSI) (CC-VSI). (**b**) Operating with a grid support and constitution inverter (GSCI) (VC-VSI).

It can be concluded that the SVS and SFS methods reduce the NDZ, generating considerable variations in the voltage or the frequency to increase islanding detection capabilities. Nevertheless, it is essential to remark that the $k_\omega$ and $k_U$ factors gains are significantly higher when applied to VC-VSIs instead of CC-VSIs. The higher the $k_\omega$ and $k_U$ gain factors are, the higher the perturbation under grid-connected operation is required. This fact affects the quality indices of the electric magnitudes. Thus, conflicts with grid regulations can appear, meaning that maybe it is not a valid method for certified installations.

### 4.3. Active Method-Impedance Measurement Type

The aim of any IM active method is to determine the impedance measurement at the PCC. In order to achieve this objective, Ohm's law can be applied, then the voltage and current values of the PCC are needed. However, the proportionality between voltage and current is only maintained if there are not any active sources participating. In this sense, to decouple the impedance value from any voltage source in the system, the voltage and current information should be obtained at a different frequency from the grid-rated one. Thus, a little disturbance is added to produce a non-characteristic harmonic current in the grid. The same harmonic voltage value is obtained from the PCC. The key challenge is to select a suitable frequency that makes the detection feasible without exceeding the maximum total harmonic distortion level.

Analyzing the transfer function of the PCC impedance, the asymptotic magnitude Bode diagram for the grid-connected and grid-disconnected operation modes has been parametrized. Figure 8 represents the transfer function where the open loop block is the resonant load impedance and the feedback block is the grid equivalent series *RL* (resistive-inductive) model admittance, both represented also in Figure 1.

The following section introduces the detection frequency bandwidth (DFBW) as a valid mechanism to explore the mentioned desired suitable frequency making IM valid and optimal.

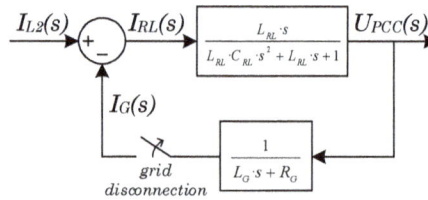

**Figure 8.** Bloc diagram of the PCC impedance transfer function.

The DFBW for IM Applicability

This paper presents a new analysis tool called DFBW that allows identifying the applicability limits of IM anti-islanding methods. The transfer function represented in Figure 8 is analyzed in detail to define the introduced concept of DFBW.

On the one hand, for the grid-connected case, the impedance Bode diagram has three characteristic angular frequencies:

-   The resonant frequency between the grid resistance and the resonant load inductance $\omega_1 = R_G/L_{RL}$.
-   The resonant frequency between the grid resistance and inductance $\omega_2 = R_G/L_G$.
-   The resonant frequency between the grid inductance and the resonant load capacitance $\omega_4 = 1/\sqrt{L_G \cdot C_{RL}}$.

On the other hand, for the grid-disconnected case, the Bode diagram presents a resonance between the resonant load inductance and capacitance at $\omega_3 = 1/\sqrt{L_{RL} \cdot C_{RL}}$.

If the different asymptotes of the transfer function depicted in Figure 8 are drawn, considering $\omega_1$–$\omega_4$, the area between the grid and the local load impedance Bode diagram allows defining a DFBW, as shown in Figure 9. This is a bandwidth that goes from the resonance between the grid resistance and the local load inductance $\omega_1$ to the resonance between the grid inductance and the local load capacitance $\omega_4$, and it can be expressed as:

$$DFBW = \frac{L_{RL}}{R_G \cdot \sqrt{L_G \cdot C_{RL}}} \tag{9}$$

where $R_G$ and $L_G$ are the resistance and the inductance of the grid.

Nevertheless, harmonic injection is usually used in an IM. Hence, the useful bandwidth loses the sub-harmonics area and starts on the local load resonance, which matches with the fundamental grid frequency $\omega_3$. Then, it is possible to redefine a new DFBW as:

$$DFBW' = \sqrt{\frac{L_{RL}}{L_G}} \tag{10}$$

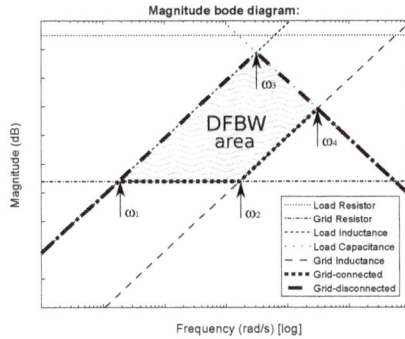

**Figure 9.** Conceptual asymptotic magnitude Bode diagram of the PCC impedance. DFBW, detection frequency bandwidth.

The asymptotes shown in Figure 9 are directly related to the weakness of the grid and to the quality factor $q$ of the resonant load. Figures 10 and 11 illustrate this behavior. In order to facilitate the analysis of how the asymptotes are shifted, a resonant load set at $R_{RL} = 1.76\ \Omega$, $C_{RL} = 1.8$ mF, and $L_{RL} = 5.6$ mH will be considered, i.e., a 30-kW/phase resonant load. For the strong grid case, it is assumed $L_G = 30$ mH and $R_G = 5$ mΩ, while for the weak grid case, $L_G = 300$ mH and $R_G = 50$ mΩ are taken into account.

Figure 10a shows the Bode diagrams of Figure 8 for both connected and disconnected scenarios assuming the mentioned 30-kW/phase resonant load, the strong grid case, and a quality factor $q$ set to one (the value suggested in [7,11,12]). It can be observed that the most sensitive change is near $\omega_3$ (314.16 rad/s for the applied case). This responds to the assumed fundamental grid frequency in the paper. Around the utility angular frequency, it is expected that the grid-disconnected impedance will be higher than the grid-connected one in terms of gain. For higher frequencies, the detection could be unattainable. This is demonstrated in Figure 10b,c, where it can be seen how the PCC impedance does not change at 350 Hz (≈2200 rad/s). This example allows deducing that the optimal multiple of the rated frequency for detection is 100 Hz (628.32 rad/s).

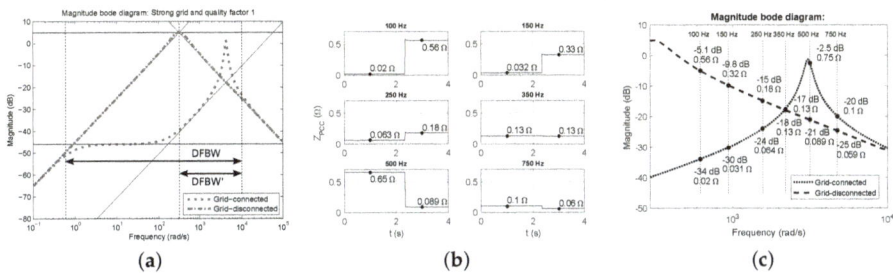

**Figure 10.** Frequency and impedance change analysis of the PCC impedance transfer function for a strong grid scenario and 30-kW/phase with $q = 1$ resonant load. (**a**) Bode diagram (magnitude). (**b**) Simulations of a PCC impedance shift after a grid disconnection. (**c**) PCC simulation test impedances on the Bode diagram.

In addition, it should be noted that the DFBW not only evolves with the weakness or strength of the mains, but also with $q$. As can be observed in Figure 11a, the DFBW is reduced for weak grids. This is due to the vertical displacement (upwards) of the grid resistance asymptote and the horizontal displacement (towards the left) of the grid inductance asymptote. Besides, considering Figure 11b, it can seen that the higher the resonant load quality factor, the narrower the DFBW. This is because of displacement towards the right of the load inductance asymptote and towards the left shift of the load capacitor asymptote.

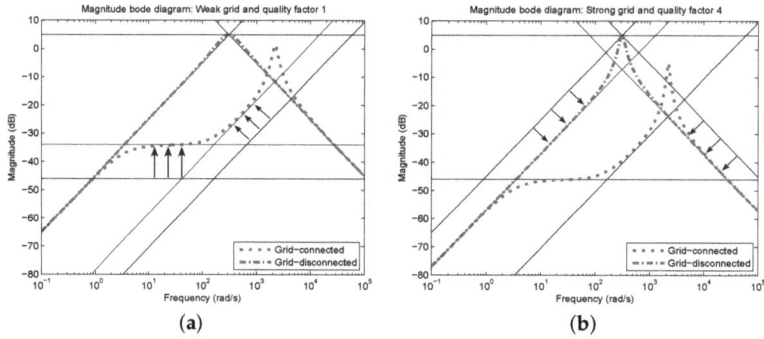

**Figure 11.** Bode diagram of the impedance at the PCC (magnitude plot). (**a**) Weak grid effect. (**b**) *q* factor effect (increase value).

*4.4. Partial Conclusions*

As a conclusion of this section, Table 3 summarizes per each previously-explored algorithm if it is valid as AIDM for GSCI and details the possible interactions with this inverter type. Note that even in the case of using an AM-IM as AIDM, it is not always valid and is highly dependent on the type of grid, the resonant load quality factor, and mainly, on the frequency used for the detection.

**Table 3.** Anti-islanding detection method (AIDM) versus grid support and constitution inverter (GSCI) interactions.

| AIDM Category | AIDM Type | GSCI Interactions Issues | Valid? |
|---|---|---|---|
| Passive method (PM) | Voltage (*V*) Frequency (*f*) | *V* and *f* indirectly controlled by the inverter | × |
| Active method-positive feedback (AM-PF) | Slip mode frequency shift (SMS) Active frequency drift (AFD) | Must be implemented in current | × |
| | Sandia voltage shift (SVS) Sandia frequency shift (SFS) | Too slow and not suitable proportional gains | × |
| Active method-impedance measurement (AM-IM) | Low frequency | Out of detection frequency bandwidth (DFBW) | × |
| | High frequency | Islanding detection feasible | ok |

**5. Improved Performance of Harmonic Injection by the Phase Perturbation Method**

The AI method suggested in this paper is based on a combination of some existing solutions [29,38,39], which have been selected and adapted to fit in an experimental platform with low availability of computational time. These high computational requirements are typical in the case of GSCI due to the extra control needs (more control loops) and extra secondary/tertiary loops to enhance droop control [4,5,9,10]. This situation makes it especially important not to overload the inverter source code with excessive computational burden to add an AIDM. The method, presented in Figure 12, is based on different steps classified into sections for better understanding.

It should be remarked that the main idea of the proposal is to detect a change in terms of impedance, and not to be accurate in the exact value of the PCC impedance. Thus, the improvement is aligned with reducing computation burden and sharing calculations at different rates.

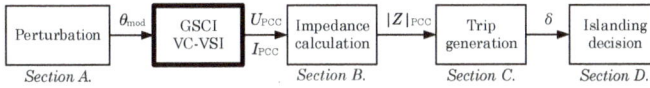

**Figure 12.** Impedance measurement operation diagram.

### 5.1. The Perturbation Injection (Section A)

As has been demonstrated in Section 4.3, the proper frequency for a 50-Hz rated grid is close to 100 Hz. Based on the method explained in [29], the phase reference of the voltage on the AC capacitor $U_C^*$ to inject a 100-Hz disturbance can be described by:

$$\theta_{mod}^* = \theta^* + k \cdot \cos \theta^* \tag{11}$$

where $k$ is an amplifying gain of the disturbance and $\theta^*$ is the phase of the referenced voltage given by the droop control. In [29], it is demonstrated that the consequence of adding this perturbation corresponds to a second harmonic injection, but without changing the zero crossing position and keeping the maximum and minimum values as the original sinusoidal waveform, as has been magnified in Figure 13. It should be highlighted that this perturbation method requires a sum, a product, and a cosine. However, the cosine argument is directly $\theta^*$ and avoids the inner product $(2 \cdot \pi \cdot \omega_r)$, where $\omega_r$ is the rated angular frequency of the grid.

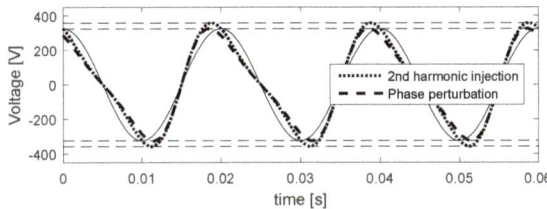

**Figure 13.** Magnified effect of phase perturbation distortion with $k = 0.5$.

### 5.2. Impedance Calculation (Section B)

In [29], the island detection method is based on variations of the quadrature synchronous reference voltage considering a symmetrical and balanced system. The methodology presented in this paper is able to detect the island in a generic unbalanced system, as in [38,39], where the detection is based on impedance measurement.

In the case of a four-wire microgrid inverter intended to operate with different power references per phase, an individual voltage measurement of each phase is needed, i.e., three degrees of freedom are being considered. Consequently, the voltage and current 100-Hz amplitude per phase must be obtained. As shown in Figure 14, the discrete Fourier transform (DTF) is used. In this case, a DFT is a mathematics calculation that requires high computational burden. However, this is implicit in any IM-based AIDM. As AIDM, the IM does not require being accurate in value, but it should be precise in value change. Thus, it is proposed in Section 5.5 to calculate the DFT at a lower frequency than that used for control purposes.

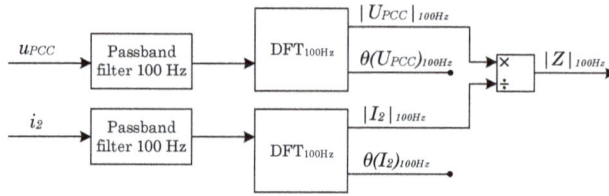

**Figure 14.** Impedance measurement diagram.

### 5.3. Trip Generation (Section C)

Sections A and B are intended to obtain an impedance value at PCC. The accurate value of the impedance itself is not the aim of the strategy. For this reason, the present section looks for creating a trigger signal indicating that a change in the impedance has occurred. To do this, a $\delta$ pulse will be created. This $\delta$ pulse contains the information of a grid-connected to grid-disconnected transition. The $\delta$ signal generation is obtained according to [29]. As a difference with respect to [29], instead of using a delay and average calculations, the $\delta$ pulse detection is replaced by filters; see Figure 15a. In this sense, it is possible to decrease the memory and the required computational time.

Figure 15b shows how the trip generation works. The first order low-pass filter, the faster one (about ms), intends to prevent erroneous island detections. Taking into account the time needed for the DFT, it should be set with a rising time close to the utility voltage period. Then, the second order filter, the slower one (about s), is designed to take off with a lower slope compared with the first order filter. In this sense, the $\delta$ pulse signal shows a fast rising behavior when a grid impedance change occurs and is slowly reduced to zero with a dynamics dominated by the second order filter response. Note that the $\delta$ has dynamic times in the order of the fast filter (ms).

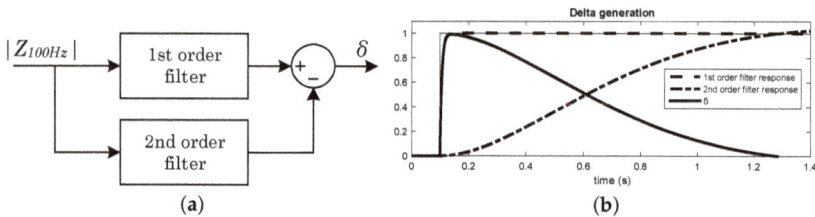

**Figure 15.** Trip generation method ($\delta$ pulse). (**a**) Generation of the $\delta$ signal. (**b**) $\delta$ pulse generation.

### 5.4. Islanding Decision (Section D)

According to all previous sections of the algorithm (A–C), at this point, it is possible to assume that the impedance at the PCC has suffered a change. However, the last step of the algorithm is presented in Figure 16 to provide a more stable decision to the islanding episode. Thus, a potential true/false grid failure flag is generated.

This final step takes the $\delta$ value from the output of Section C and contrasts it with a $\delta_{lim}$. The $\delta_{lim}$ is tuned considering the perturbation frequency and the expected DFBW. When $\delta$ exceeds $\delta_{lim}$, a counter allows shifting in time the decision to ensure a real detection.

Using an IM anti-islanding strategy, different from a PF one, the detection time is not directly proportional to the injected perturbation. The bigger the perturbation is, the more reliable the impedance measurement becomes, not the faster. In fact, the detection time is only fixed by the gap needed for the DFT calculation and the arranged delay by the decision.

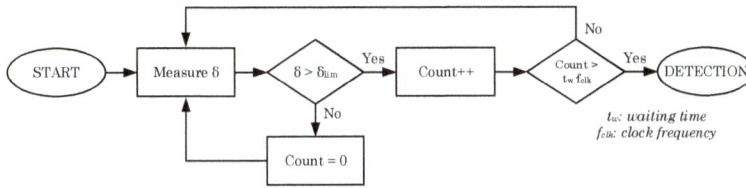

**Figure 16.** Islanding decision steps.

### 5.5. Calculation Rates Proposal

According to all previous subsections, it is proposed to apply the algorithm presented in Figure 12, but sharing the calculations at two frequency rates: one at high frequency concerning inner control loop implications and one at low frequency for the detection heavy cost functions such as the DFT, as described in Figure 17.

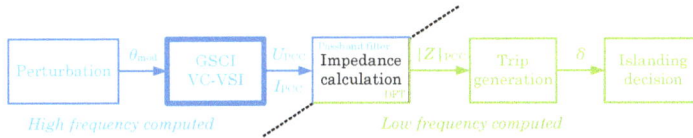

**Figure 17.** Impedance measurement operation diagram.

## 6. Experimental Results

This section presents the setup used for the validation of the improved IM developed in Section 5. It also evaluates the proposal in terms of cost under the setup used and the feasibility in time for the detection.

### 6.1. Setup

The AI algorithm has been implemented on the 90-kVA VC-VSI inverter based on three Semikron IGD-2-424 power stacks, as shown in Figure 18. The converter was a three-phase four-wire inverter, whose inner control loops (current and voltage) were based on adaptive proportional resonant controllers (PR) [30]. Each active phase was controlled independently by managing the active and reactive power set-points through an AC droop control strategy, as depicted in Figure 1a. The controller parameters and the values of the considered LCL output filter are summarized in Table 2 (same values used for simulations in Section 4.2.4).

The resonant load considered was 13.8 kW (three-phase) assuming a quality factor $q$ of 0.5 ($R_{RL} = 11.5\ \Omega$; $L_{RL} = 73$ mH; $C_{RL} = 138\ \mu$F). In Figure 19, the experimental resonant load used is shown. As can be observed, the required physical component to achieve $q$ at 0.5 are really bulky, requiring uncommon values of inductances, resulting in it being even more difficult to achieve $q$ values of one or two, as requested in [7,11,12].

The detection algorithm was implemented in a TMS320F2809 DSP from Texas Instruments. The perturbation was adjusted injecting a 3.53-$A_{rms}$ second harmonic component, which corresponds to 2.5% current harmonic distortion in grid-connected mode (when off grid, the voltage harmonic distortion was less than 2%). The islanding decision was designed to identify steps of more than 0.4 $\Omega$ holding during at least 50 ms.

**Figure 18.** Four-wire 90-kVA VC-VSI inverter.

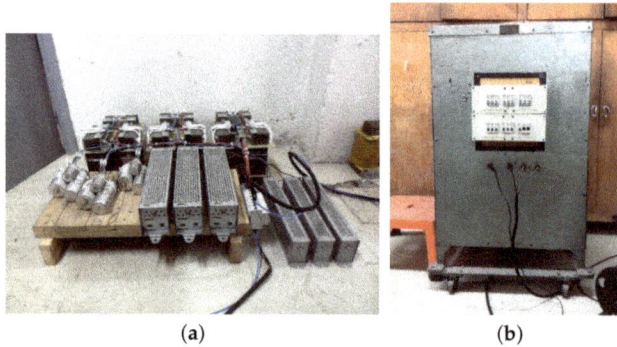

(**a**)                                          (**b**)

**Figure 19.** Resonant load picture. (**a**) Part A (inductances, capacitors, and low power resistances). (**b**) Part B (big resistive block).

### 6.2. Results

Computational Cost

As described in Section 5.5, the calculations required to apply the proposed IM as an AIDM were split at two frequency rates. The high frequency calculations were executed every 8 kHz (125 μs) and the low frequency ones at 1 kHz (1 ms). In Figure 20, the different times required to compute each section can be seen.

The total time for the high frequency calculations represented less than 4% of the available 125 μs, and for the low frequency case, less than 0.5% of the available 1 ms.

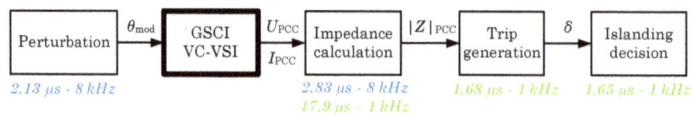

**Figure 20.** Impedance measurement operation diagram.

## 6.3. Detection Validation

This section evaluates the detection time in the experimental setup dividing the analysis of the ZPF and NZPF scenarios.

### 6.3.1. Detection under Zero Power Flow

The algorithm response under ZPF conditions, where all the local loads are consuming all the generated power, can be seen in Figure 21a,b. Figure 21a refers to a detection considering that the inverter is delivering null power and there are no local loads. The detection time was achieved in 80 ms. Figure 21b presents another ZPF scenario. In this case, the inverter is delivering 4.6 kW per phase, but all this power is used to feed a local load connected at the PCC. The detection time was achieved in about 80 ms, as well.

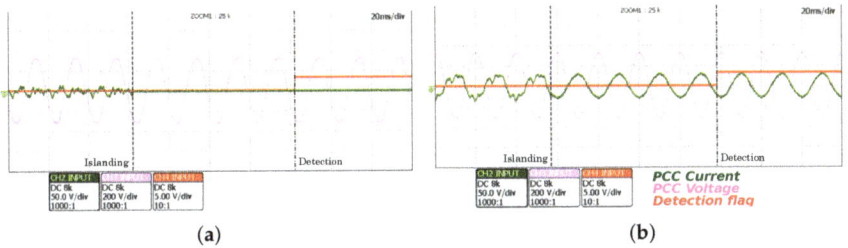

**Figure 21.** Experimental results under zero power flow. (**a**) Null power reference zero and no local loads. (**b**) Inverter delivered power matching with the local resonant load consumption (4.6-kW/phase).

### 6.3.2. Detection under Non-Zero Power Flow

In the case of considering VC-VSI, both ZPF or NZPF conditions can make the island detection difficult. As has been mentioned before, this is due to the VC-VSI being the device that maintains the voltage at the PCC and not the resonant load (as is the case when CC-VSI are considered). Figure 22a–d shows the validity of the suggested AIDM also under NZPF scenarios. Figure 22a,b illustrates the detection trigger when the inverter was delivering less power than the resonant local load required. The detection times obtained were 77 and 70 ms, respectively. Finally, Figure 22c,d shows the detection trigger when the inverter exceeded the power required to feed the local load. In these last scenarios, the detections were achieved in 64 ms and 81 ms, respectively.

Note that even considering NZPF cases, the voltage magnitudes at the PCC (voltage and frequency) were practically unaltered. This is the key point of the need not to consider AIDM based on voltage or frequency displacements.

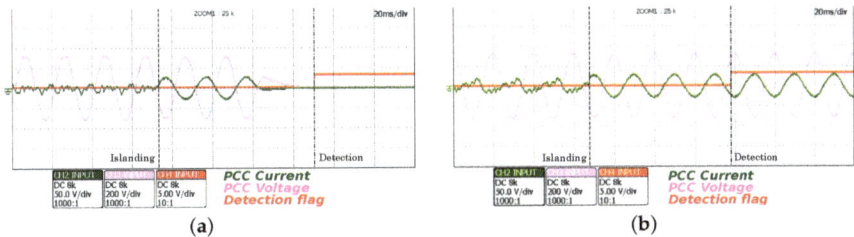

**Figure 22.** Cont.

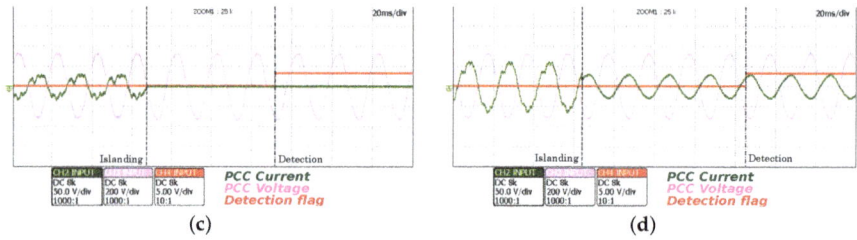

**Figure 22.** Experimental results under non-zero power flow. (**a**) Zero power injection and 4.6-kW resonant load per phase. (**b**) Two-kilowatt power injection and 4.6-kW resonant load per phase. (**c**) Five-kilowatt power injection without local load. (**d**) Ten-kilowatt power injection and 4.6-kW resonant load per phase.

## 7. Conclusions

This paper presents the difference of operation applied to the anti-islanding detection methods (AIDM) when an inverter is assumed as a voltage-controlled source (VC-VSI) or current-controlled source (CC-VSI). The existence of these two types of inverters limits the conventional resonant load concept required for island detection. Thus, the concept of zero and non-zero power flow (ZPF/NZPF) was introduced to extend the worst case for island detection for either CC-VSI or VC-VSI.

The paper also clearly exposed the problem of several (passive methods, four active methods based on positive feedback strategies and impedance measurement) AIDMs operating towards VC-VSIs, especially focused on GSCIs (inverters under AC droop control strategies widely used to parallelize several units). Moreover, the effects of using AIDM methods based on voltage and frequency displacement, passive and active (positive feedback), for which the detection is not feasible, have been presented. The analysis was extended to impedance measurement (IM) AIDM, and its limitations have been detailed. This applicability limitations' analysis was conducted to present the detection frequency bandwidth (DFBW) as a new tool that allows determining which external conditions are compatible with the detection feasibility; in other words, which kind of loads and grids (weak/strong) will be in the detection framework. In this sense, the optimal frequency to determine the impedance can be calculated.

The use of an AIDM (IM or other) in VC-VSI inverters requires a certain computational burden that usually is limited for such inverters. Thus, an improved IM method was proposed as a hybrid option based on existing methods, but focusing on the detection change and not on the impedance value accuracy. Thus, the proposed IM AIDM was split into high and low computation requirements, allowing it to be more time efficient.

The proposed IM AIDM anti-islanding method has been designed, simulated, and tested in a 90-kVA four-wire three-phase inverter. The selection of the optimal frequency to be injected for the detection method has been selected using the proposed DFBW concept. All the conducted tests have concluded that the detection strategy can be implemented with low computational costs; less than 4% for the high frequency calculations and less than 0.5% for the low frequency ones. Furthermore, the IM AIDM applied is able to generate an island detection flag below 100 ms. The detection time obtained with this method is below the minimum detection time fixed by standards of 160 ms. It is important to remark that for an IM, the consideration of a resonant load under zero power flow conditions at the PCC does not add time to the detection. Thus, for the present IM proposal, the strictest time requirements have been obtained over ZPF or NZPF situations, and CC-VSI or VC-VSI became valid, general, and efficient AIDM with a reduced computational cost.

**Author Contributions:** Conceptualization, M.L.-M. and D.H.-P.; methodology, M.L.-M. and D.H.-P.; software, M.L.-M.; validation, M.L.-M. and D.H.-P.; formal analysis, M.L.-M. and D.H.-P.; investigation, M.L.-M., D.H.-P., C.C.-A., R.V.-R., and D.M.-M.; resources, D.M.-M.; writing, original draft preparation, M.L.-M. and D.H.-P.; writing, review and editing, M.L.-M., D.H.-P., and C.C.-A.; visualization, M.L.-M., D.H.-P., C.C.-A., and R.V.-R.; supervision, D.H.-P., R.V.-R., and D.M.-M.

**Funding:** This research received no external funding.

**Conflicts of Interest:** The authors declare no conflict of interest.

## References

1. Paris: REN21 Secretariat. *Renewables 2017 Global Status Report*; Technical Report; REN21: Paris, France, 2017. ISBN 978-3-9818107-6-9. Available online: http://www.ren21.net/wp-content/uploads/2017/06/17-8399_GSR_2017_Full_Report_0621_Opt.pdf (accessed on 13 March 2019).

2. Ramezani, M.; Li, S.; Sun, Y. Combining droop and direct current vector control for control of parallel inverters in microgrid. *IET Renew. Power Gener.* **2017**, *11*, 107–114. [CrossRef]

3. Trivedi, A.; Singh, M. Repetitive Controller for VSIs in Droop-Based AC-Microgrid. *IEEE Trans. Power Electron.* **2017**, *32*, 6595–6604. [CrossRef]

4. Heredero-Peris, D.; Chillón-Antón, C.; Pagès-Giménez, M.; Montesinos-Miracle, D.; Santamaría, M.; Rivas, D.; Aguado, M. An Enhancing Fault Current Limitation Hybrid Droop/V-f Control for Grid-Tied Four-Wire Inverters in AC Microgrids. *Appl. Sci.* **2018**, *8*, 1725. [CrossRef]

5. De Brabandere, K. Voltage and Frequency Droop Control in Low Voltage Grids by Distributed Generators with Inverter Front-End. Ph.D. Thesis, Katholieke Universiteit Leuven, Leuven, Belgium, 2006.

6. Xiao, W.; Elnosh, A.; Khadkikar, V.; Zeineldin, H. Overview of maximum power point tracking technologies for photovoltaic power systems. In Proceedings of the IECON 2011—37th Annual Conference of the IEEE Industrial Electronics Society, Melbourne, VIC, Australia, 7–10 November 2011; pp. 3900–3905. [CrossRef]

7. VDE-AR-N 4105:2011-08 Power Generation Systems Connected to the Low-Voltage Distribution Network. 2011. Available online: https://www.vde-verlag.de/standards/0105029/vde-ar-n-4105-anwendungsregel-2011-08.html (accessed on 13 March 2019).

8. Li, C.; Savaghebi, M.; Vasquez, J.C.; Guerrero, J.M. Multiagent based distributed control for operation cost minimization of droop controlled AC microgrid using incremental cost consensus. In Proceedings of the 2015 17th European Conference on Power Electronics and Applications, EPE-ECCE Europe 2015, Geneva, Switzerland, 8–10 September 2015. [CrossRef]

9. Li, C.; Coelho, E.A.A.; Savaghebi, M.; Vasquez, J.C.; Guerrero, J.M. Active power regulation based on droop for AC microgrid. In Proceedings of the 2015 IEEE 10th International Symposium on Diagnostics for Electrical Machines, Power Electronics and Drives (SDEMPED), Guarda, Portugal, 1–4 September 2015; pp. 508–512. [CrossRef]

10. Zhong, Q.; Zeng, Y. Universal Droop Control of Inverters with Different Types of Output Impedance. *IEEE Access* **2016**, *4*, 702–712. [CrossRef]

11. *IEEE 1547 Standard for Interconnecting Distributed Resources with Electric Power Systems*; IEEE: Piscataway, NJ, USA, 2003. [CrossRef]

12. IEC 61727 Photovoltaic (PV)—Characteristics of the Utility Interface. 2004. Available online: https://standards.globalspec.com/std/365170/iec-61727 (accessed on 13 March 2019).

13. Ravichandran, A.; Malysz, P.; Sirouspour, S.; Emadi, A. The critical role of microgrids in transition to a smarter grid: A technical review. In Proceedings of the 2013 IEEE Transportation Electrification Conference and Expo: Components, Systems, and Power Electronics—From Technology to Business and Public Policy, ITEC 2013, Detroit, MI, USA, 16–19 June 2013. [CrossRef]

14. Hossain, M.J.; Mahmud, M.A.; Pota, H.R.; Mithulananthan, N.; Bansal, R.C. Distributed control scheme to regulate power flow and minimize interactions in multiple microgrids. In Proceedings of the IEEE Power and Energy Society General Meeting, National Harbor, MD, USA, 27–31 July 2014; pp. 1–5. [CrossRef]

15. Murthy Balijepalli, V.S.K.; Khaparde, S.A.; Gupta, R.P.; Pradeep, Y. SmartGrid initiatives and power market in India. In Proceedings of the IEEE PES General Meeting, PES 2010, Providence, RI, USA, 25–29 July 2010; pp. 1–7. [CrossRef]

16. Yu, D.; Lian, B.; Dunn, R. Using Control Methods to Model Energy Hub Systems. In Proceedings of the Power Engineering Conference (UPEC), Cluj-Napoca, Romania, 2–5 September 2014; pp. 1–4.

17. Girbau-Llistuella, F.; Rodriguez-Bernuz, J.M.; Prieto-Araujo, E.; Sumper, A. Experimental Validation of a Single Phase Intelligent Power Router. In Proceedings of the Innovative Smart Grid Technologies Conference Europe (ISGT-Europe), 2014 IEEE PES, Istanbul, Turkey, 12–15 October 2014; pp. 1–6.

18. Monyei, C.G.; Fakolujo, O.A.; Bolanle, M.K. Virtual power plants: Stochastic techniques for effective costing and participation in a decentralized electricity network in Nigeria. In Proceedings of the 2nd International Conference on Emerging and Sustainable Technologies for Power and ICT in a Developing Society, IEEE NIGERCON 2013—Proceedings, Owerri, Nigeria, 14–16 November 2013; pp. 374–377. [CrossRef]

19. Tirumala, R.; Mohan, N.; Henze, C. Seamless transfer of grid-connected PWM inverters between utility-interactive and stand-alone modes. In Proceedings of the APEC, Seventeenth Annual IEEE Applied Power Electronics Conference and Exposition (Cat. No. 02CH37335), Dallas, TX, USA, 10–14 March 2002; Volume 2, pp. 1081–1086. [CrossRef]

20. Serban, E.; Pondiche, C.; Ordonez, M. Islanding Detection Search Sequence for Distributed Power Generators Under AC Grid Faults. *IEEE Trans. Power Electron.* **2015**, *30*, 3106–3121. [CrossRef]

21. Guo, Z. A harmonic current injection control scheme for active islanding detection of grid-connected inverters. In Proceedings of the 2015 IEEE International Telecommunications Energy Conference (INTELEC), Osaka, Japan, 18–22 October 2015; pp. 1–5. [CrossRef]

22. Das, P.P.; Chattopadhyay, S. A Voltage-Independent Islanding Detection Method and Low-Voltage Ride Through of a Two-Stage PV Inverter. *IEEE Trans. Ind. Appl.* **2018**, *54*, 2773–2783. [CrossRef]

23. Murugesan, S.; Murali, V.; Daniel, S.A. Hybrid Analyzing Technique for Active Islanding Detection Based ond-Axis Current Injection. *IEEE Syst. J.* **2018**, *12*, 3608–3617. [CrossRef]

24. Park, S.; Kwon, M.; Choi, S. Reactive Power P amp;O Anti-Islanding Method for a Grid-Connected Inverter With Critical Load. *IEEE Trans. Power Electron.* **2019**, *34*, 204–212. [CrossRef]

25. Xiao, H.F.; Fang, Z.; Xu, D.; Venkatesh, B.; Singh, B. Anti-Islanding Protection Relay for Medium Voltage Feeder With Multiple Distributed Generators. *IEEE Trans. Ind. Electron.* **2017**, *64*, 7874–7885. [CrossRef]

26. Pouryekta, A.; Ramachandaramurthy, V.K.; Mithulananthan, N.; Arulampalam, A. Islanding Detection and Enhancement of Microgrid Performance. *IEEE Syst. J.* **2018**, *12*, 3131–3141. [CrossRef]

27. Llonch-Masachs, M.; Heredero-Peris, D.; Montesinos-Miracle, D. An anti-islanding method for voltage-controlled VSI. In Proceedings of the 2015 17th European Conference on Power Electronics and Applications (EPE'15 ECCE-Europe), Geneva, Switzerland, 8–10 September 2015; pp. 1–10. [CrossRef]

28. Chillon-Anton, C.; Llonch-Masachs, M.; Heredero-Peris, D.; Pages-Gimenez, M.; Montesinos-Miracle, D. Assisting passive anti-islanding proposal for Voltage-Controlled Voltage-Source-Inverters. In Proceedings of the PCIM Europe 2018; International Exhibition and Conference for Power Electronics, Intelligent Motion, Renewable Energy and Energy Management, Nuremberg, Germany, 5–7 June 2018; pp. 1–8.

29. Ciobotaru, M.; Agelidis, V.G.; Teodorescu, R.; Blaabjerg, F. Accurate and less-disturbing active antiislanding method based on pll for grid-connected converters. *IEEE Trans. Power Electron.* **2008**, *25*, 1576–1584. [CrossRef]

30. Heredero-Peris, D.; Pages-Gimenez, M.; Montesinos-Miracle, D. Inverter design for four-wire microgrids. In Proceedings of the 2015 17th European Conference on Power Electronics and Applications, EPE-ECCE Europe 2015, Geneva, Switzerland, 8–10 September 2015. [CrossRef]

31. Teodorescu, R.; Liserre, M.; Rodriguez, P. *Grid Converters for Photovoltaic and Photovoltaic and Wind Power Systems*; John Wiley & Sons, Ltd.: Hoboken, NJ, USA, 2011; ISBN 9780470057513.

32. Bower, W.; Ropp, M. Evaluation of Islanding Detection Methods for Utility-Interactive Inverters in Photovoltaic Systems; Sandia Report 2002. 2002. Available online: http://www.iea-pvps.org/index.php?id=9&eID=dam_frontend_push&docID=386 (accessed on 13 March 2019).

33. Singam, B.; Hui, L.Y. Assessing SMS and PJD schemes of anti-islanding with varying quality factor. In Proceedings of the First International Power and Energy Conference, (PECon 2006) Proceedings, Putra Jaya, Malaysia, 28–29 November 2006; pp. 196–201. [CrossRef]

34. Liu, F.; Kang, Y.; Zhang, Y.; Duan, S.; Lin, X. Improved SMS islanding detection method for grid-connected converters. *IET Renew. Power Gener.* **2010**, *4*, 36–42. [CrossRef]

35. Ye, Z.; Walling, R.; Garces, L.; Zhou, R.; Li, L.; Wang, T. *Study and Development of Anti-Islanding Control for Grid-Connected Inverters*; National Renewable Energy Laboratory: Lakewood, CO, USA, 2004; p. 82.

36. Hopewell, P.; Jenkins, N.; Cross, A. Loss-of-mains detection for small generators. *IEE Proc. Electr. Power Appl.* **1996**, *143*, 225–230. [CrossRef]

37. Kane, P.O.; Fox, B.; Mains, L.O.F. Loss of Mains Detection for Embedded Generation By System Impedance Monitoring. In Proceedings of the Sixth International Conference on Developments in Power System Protection, Nottingham, UK, 25–27 March 1997; pp. 95–98.

38. Ciobotaru, M.; Teodorescu, R.; Blaabjerg, F. On-line grid impedance estimation based on harmonic injection for grid-connected PV inverter. In Proceedings of the IEEE International Symposium on Industrial Electronics, Vigo, Spain, 4–7 June 2007; pp. 2437–2442. [CrossRef]

39. Asiminoaei, L.; Teodorescu, R.; Blaabjerg, F.; Borup, U. A new method of on-line grid impedance estimation for PV inverter. In Proceedings of the Nineteenth Annual IEEE Applied Power Electronics Conference and Exposition, APEC'04, Anaheim, CA, USA, 22–26 February 2004; Volume 3. [CrossRef]

40. Asiminoaei, L.; Teodorescu, R.; Blaabjerg, F.; Borup, U. Implementation and test of on-Line embedded grid impedance estimation for PV-inverters. In Proceedings of the PESC Record—IEEE Annual Power Electronics Specialists Conference, Aachen, Germany, 20–25 June 2004; Volume 4, pp. 3095–3101. [CrossRef]

41. Asiminoaei, L.; Teodorescu, R.; Blaabjerg, F.; Borup, U. A digital controlled PV-inverter with grid impedance estimation for ENS detection. *IEEE Trans. Power Electron.* **2005**, *20*, 1480–1490. [CrossRef]

42. Liu, S.; Li, Y.; Ji, F.; Xiang, J. Islanding detection method based on system identification. *IET Power Electron.* **2016**, *9*, 2095–2102. [CrossRef]

*applied sciences*

MDPI

*Review*

# Contribution of Smart Cities to the Energy Sustainability of the Binomial between City and Country

**Manuel Villa-Arrieta [1] and Andreas Sumper [2,\*]**

[1] Escola Tècnica Superior d'Enginyeria Industrial de Barcelona (ETSEIB), Universitat Politècnica de Catalunya (UPC), Av. Diagonal, 647, 08028 Barcelona, Spain

[2] Centre d'Innovació Tecnològica en Convertidors Estàtics i Accionaments (CITCEA-UPC), Escola Tècnica Superior d'Enginyeria Industrial de Barcelona (ETSEIB), Universitat Politècnica de Catalunya (UPC), Av. Diagonal, 647, Pl. 2, 08028 Barcelona, Spain

\* Correspondence: andreas.sumper@upc.edu; Tel.: +34-93-401-6727

Received: 30 June 2019; Accepted: 6 August 2019; Published: 8 August 2019

**Featured Application: This work analyzes the contribution of energy self-sufficiency in cities (smart cities) to energy security and the environmental sustainability of the countries.**

**Abstract:** Cities are at the center of the transition to a decarbonized economy. The high consumption of electricity in these urban areas causes them to be the main focus of greenhouse gas emissions. However, they have a high margin of capacity to increase energy efficiency and local energy generation. Along these lines, the smart urban management model has been proposed as a solution to the unsustainability of cities. Due to the global trend of population concentration in urban areas, cities tend to be representative of the population, energy consumption, and energy sustainability of their countries. Based on this hypothesis, this paper studied the relationship between the smart city model and the concept of energy sustainability. First, the research analyzed the relationship between urban population growth and energy sustainability; and then the self-consumption capacity of photovoltaic electricity of the main cities of the countries classified in the energy sustainability indicator (Energy Trilemma Index 2017) of the World Energy Council was analyzed. According to the results, the scope of action for self-consumption of photovoltaic electricity is broad and cities have the capacity to contribute significantly to the energy sustainability of their countries. Following the approach of other authors, the development of energy sustainability objectives and the installation of smart systems in distribution grids must be aligned with national objectives.

**Keywords:** energy sustainability; smart cities; PV self-consumption; Energy Trilemma Index; energy security; environmental sustainability

## 1. Introduction

Covering the current needs of humanity without jeopardizing the coverage of the needs of future generations is the definition established for sustainable development [1]. Thus, the concept of energy sustainability is used to address the coverage of the energy demand of society. However, this process has not been sustainable because the development and economic growth of humanity has been based on the use of energy resources of fossil origin. The transformation of these resources to final energy has had an impact on the emission of greenhouse gases (GHG), the precursors of climate change.

This final energy, mainly in the form of electricity, has been mainly destined to meet the energy demand of the urban population. As seen in Figure 1 (which includes data after the 1973 oil crisis), electricity consumption has a greater correlation with the growth of the urban population than with the growth of the total population. As a further handicap, by 2040 the global demand for energy will have risen by more than a quarter due to the increase of people residing in urban areas in developing economies to 1.7 billion [2]; by 2050, 75% of the population will live in the cities [3].

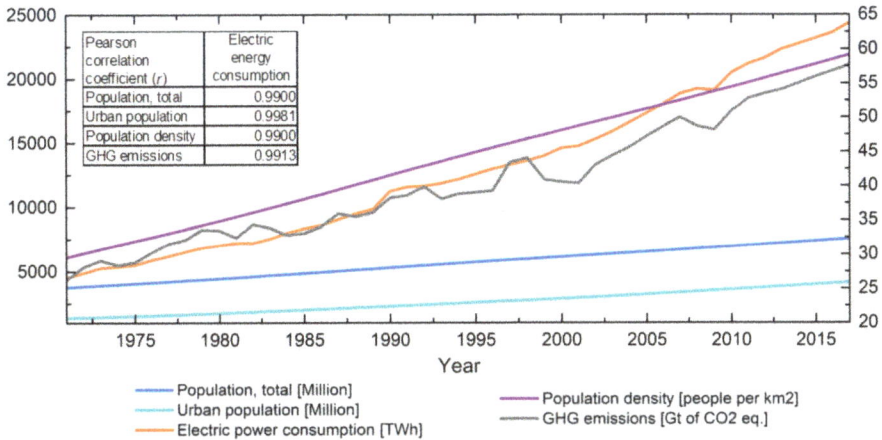

| Pearson correlation coefficient (*r*) | Electric energy consumption |
|---|---|
| Population, total | 0.9900 |
| Urban population | 0.9981 |
| Population density | 0.9900 |
| GHG emissions | 0.9913 |

— Population, total [Million]
— Urban population [Million]
— Electric power consumption [TWh]
— Population density [people per km2]
— GHG emissions [Gt of CO2 eq.]

**Figure 1.** Correlation between urban population and electric energy consumption. Source: Created by authors using data from [4].

Because energy consumption is transversal to economic activity, mitigation and adaptation measures towards climate change are marked by a process of the transformation of the current economic model to one decarbonized. This process is called energy transition and is based on increasing energy efficiency, electrification, and the use of renewable energy resources [5]. At the scale of the cities there is talk of urban energy transition (UET) to the use of urban renewable energy resources and the reduction of the energy consumption of buildings [6]. Compared to industry, transport, and other energy end-use consumption sectors (according to the New Policies Scenario of the World Energy Outlook 2018, in 2017 the final energy consumption was 9696 Mtoe and was distributed as follows: 3265 Mtoe in the industry (29.45%), 2794 Mtoe in transport (28.82%), 3047 Mtoe in buildings (31.43%), and 999 Mtoe (10.3%) in other energy end-use consumption sectors), world energy consumption is concentrated in buildings [2]. Thus, the greatest potential for energy saving is in these constructions [6].

In this regard, the strategies of the UET seeks to make the operation of the electrical system more flexible through the empowerment of the consumer in the management of demand and the generation of urban energy through the energy self-consumption in buildings. Making the electrical system more flexible involves an optimal combination of demand and renewable energy generation. However, the main sources of this type of energy, such as solar and wind, are intermittent: Periods of higher renewable generation generally usually occur during periods of lower residential energy demand. Therefore, to take full advantage of these energy sources, it is necessary to implement mechanisms that optimize the operation of the distribution system in cities.

A strategy aimed at addressing competitiveness and confronting the efficiency and limitation of the economic and natural resources of the urban areas is smart city model [7,8]. Although this strategy surfaced during the last decade to address the limitation of energy resources in cities and the inherent production of GHG, today it covers other fields that include the provision of urban services, and governance, knowledge, and behavior of citizens [9]. Technically, smart cities are cities with a high degree of penetration of Information and Communication Technologies (ICT) to create synergies [10] between technological components and economic agents.

Due to the global trend of population concentration in these urban areas, cities tend to be representative of the population, energy consumption, and energy sustainability of their countries.

Based on this hypothesis, this paper analyzes the theoretical contribution that the smart city model can have on the energy sustainability of countries. In this context, from theoretical and empirical approaches and with conceptual results, this research seeks to contribute to the study of UET.

In this regard, this paper provides a description of the smart city model and the concept of energy sustainability in Section 2, in order to analyze the common elements between them in Section 3. Finally, in Section 4, the paper analyzes the effect of the concentration of the urban population and the use of solar resources in cities in terms of their countries' energy sustainability.

## 2. Concepts Review

### 2.1. Energy Sustainability

According to the World Energy Council (WEC), the definition of energy sustainability is based on three core dimensions: energy security, energy equity, and environmental sustainability. Together, they make up a "trilemma" and achieving high performance in all three dimensions entails complex, interwoven links between public and private actors, governments and regulators, economic and social factors, national resources, environmental concerns, and individual behaviors [11]. In order to evaluate the performance of this energy sustainability trilemma for countries, the WEC uses the Energy Trilemma Index.

The index is a complete, rigorous, and widely-recognized indicator for decision-making on energy policy [12]. In its calculation, 75% corresponds to the analysis of the energy performance of the countries, where energy security, energy equity, and environmental sustainability are evaluated. The remaining 25% corresponds to the analysis of contextual performance. Generally speaking, a country is awarded a better rating and position in the ranking when it demonstrates a better energy and contextual performance. From now on, to talk about energy sustainability in this paper, we will only refer to energy performance, and the following acronyms will be used: ETI to talk about the Energy Trilemma Index; ESR to talk about energy security; EQR to talk about energy equity; and EVR to talk about environmental sustainability.

Energy security is the effective management of primary energy supply from domestic and external sources, reliability of energy infrastructure, and ability of energy providers to meet current and future demand; energy equity is the accessibility and affordability of energy supply across the entire population; and environmental sustainability is achievement of supply- and demand-side energy efficiencies and development of energy supply from renewable and other low-carbon sources [11]. The contextual indicators consider the broader circumstances of energy performance, including a country's ability to provide coherent, predictable and stable policy and regulatory frameworks, initiate R&D and innovation, and attract investment [12].

ETI studies 125 countries according to region and their gross domestic product per capita (group GDP or Group-Roman_number). The regions are Europe (E), Sub-Saharan Africa (Sub-Saharan, S), Asia (A), Latin America and the Caribbean (LAC, L), the Middle East (MENA, M), and North Africa (North-America, N). The GDP is divided into Group-I (more than 33,500 USD), Group-II (between 14,300 and 33,500 USD), Group-III (between 6000 and 14,300 USD), and Group-IV (less than 6000 USD). In the context of this paper, the results in 2017 of this index were organized in Figure 2. Group-I does not contain any countries form the LAC or Sub-Saharan regions and Group-IV does not contain countries in MENA, North America. Group-II is the only group with has countries from all six regions of the study.

The ETI also has an interactive tool, the pathway calculator, that can be used to determine what is necessary to improve the ranking position and understand the impact of policymaking on achieving a sustainable energy future [13]. Figure 3 presents the indicators of this virtual tool. According to this tool, basically a country obtains a better energy sustainability by diversifying its mix of primary energy sources, diversifying its mix of electricity generation, reducing energy imports, and reducing its $CO_2$ and other GHG emissions.

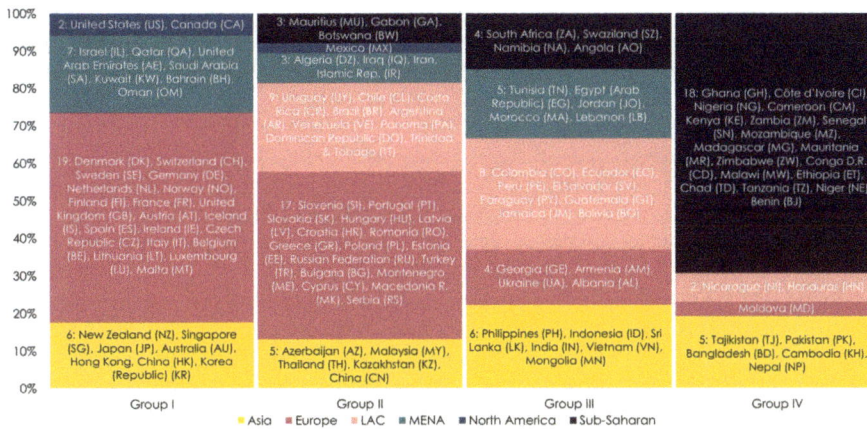

**Figure 2.** Distribution of the 125 countries of the ETI (Energy Trilemma Index) 2017 by regions and GDP groups per capita. Notes: Regions are divided (greatest to least presence) into Europe with 41 countries (32.8%), Sub-Saharan 25 (29%), Asia 22 (17.6%), LAC 19 (15.2%), MENA 15 (12%) and North-America 3 (2.4%). The GDP groups are divided as: Group-I has 34 countries (27.2%), Group-II has 38 (30.4%), Group-III has 27 (21.6%), and Group-IV has 26 (20.8%). Source: Created by authors using data from [12].

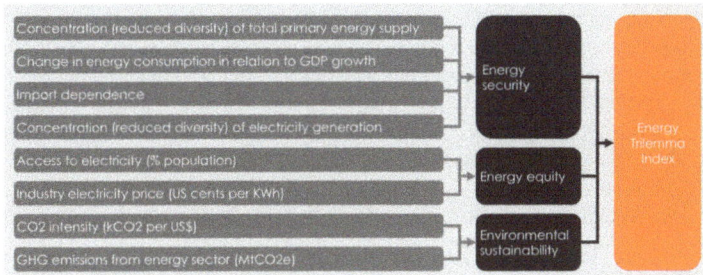

**Figure 3.** Structure of the indicators of the pathway calculator of the WEC (World Energy Council). Source: Created by authors using data from [13].

## 2.2. Smart City Model

The SMART criterion emerged in the 1980s and is an acronym of the objectives of a business management proposal: specific, measurable, achievable, relevant, time-bound [14]. Later these criteria became a concept which was applied to different disciplines of study, with the intention of improving results in terms of increased efficiency in resource management by coordinating information between different systems. From an organizational point of view, information control in cities would allow cross-communication between the actors involved in the management of cities and citizens [15]. Thus, through the advancement of ICT, the coordination of information between systems improves the SMART management approach towards its application in more complex systems such as cities [16–20].

From an integral and holistic point of view, the smart city concept brings together a series of strategic proposals, which seeks for the development and future growth of cities in order to make them competitive, sustainable, and able to offer a high quality of life to urban population [21–23]. With the control of information, the crux of these management proposals is to find out the modus operandi of the demand for urban services in order to efficiently manage the resources available: natural resources (energy, water, air), infrastructures, economic resources, people, and knowledge.

Since the early 1990s, the smart city concept has been a subject of study (a search for the term "smart city" in the Web of Science (WOS) and Scopus return the first references to it in 1991 [24] and 1997 [25]). From this time until the middle of the current decade, various authors have published their views on the definition without reaching an agreement about the scope of this concept [21,26–28]. Even so, the definitions presented included proposals for the working taxonomy of this type of city, such as in [26], evaluation and classification methodologies as in [29,30], planning in [31], suggestions for strategies to successfully implement the smart model in cities in [28,32,33], and in the case of [34], a reference initiative for this purpose, in this case it was Barcelona. These publications also included methods to integrate technological devices into the services that are to be provided in cities, such as [35].

Working with the same concept, to date, the study of the smart city concept has been done by international (the European Union's approach to the smart city concept considers energy sustainability to be a priority on the basis of the main energy resource being energy efficiency [36,37], and the conception, development, and integration of urban energy production and use. Its objective is to improve efficiency as well as reduce energy consumption and the emission of GHG [38]) organizations [22] and tech companies [39], companies in the financial sector [40], as well as centers for economic, financial, technological, and social studies. The large number of parties involved in the defining of the smart city concept shows the magnitude of its importance in the future of cities and its ability to catch the attention of the different parties involved in urban growth and development. The ultimate aim of these parties is to work on two closely related aspects of contemporary cities: quality of life and competitiveness [41]. In being competitive, cities seek to attract greater investments and offer a higher level of quality of life to its inhabitants, and they will be able to get the best out of human capabilities, thus in turn making them increasingly competitive [38].

To reach these goals, the working sectors of smart cities can be classified into "hard" and "soft" domains, depending on the importance of ICT systems operating within them [26]. According to this classification, the Hard domain includes the sectors in which ICT help to configure cities at a technical level. These are energy grids; public lighting; natural resources and water management; waste management; environment; transport, mobility, and logistics; office and residential buildings; healthcare and public security. As for the Soft domains, the presence of ICT is limited since they are sectors that do not necessarily require real-time information, processing, and integration. These sectors include education and culture, social inclusion and welfare, public administration and (e-)government and economy.

The inclusion of ICT in the systems and technological elements makes them smart technologies [42], and examples of these are sensors, electric meters, and other elements interconnected via the Internet under the concept of the Internet of Things (IOT) [43,44]. These technologies allow interoperability between components that capture and demand signals and provide a continuous response to service management. In this sense, since the energy service is a transversal element in the efficient provision of the services of cities, the smart cities are found within the framework of the urban energy transition. Figure 4 summarizes this approach in the sense that the integration of ICT in cities allows the application of smart technologies to provide the services demanded by the urban population with the aim of improving the city's quality of life, competitiveness, and sustainability.

It should be noted that other authors pose different positions or approaches to study. Authors such as [45,46] go one step further and consider the potential of smart technologies as a solution to the problems of environmental unsustainability, which is determined by the size of the carbon footprint and the environmental and energy impact that these technologies will cause during the improvement of other processes. In addition, according to [47], cities can be made sustainable without the use of smart ICT, and smart technologies can be used in cities without contributing to sustainable development; and, these technologies can also be used for sustainable development in locations other than cities.

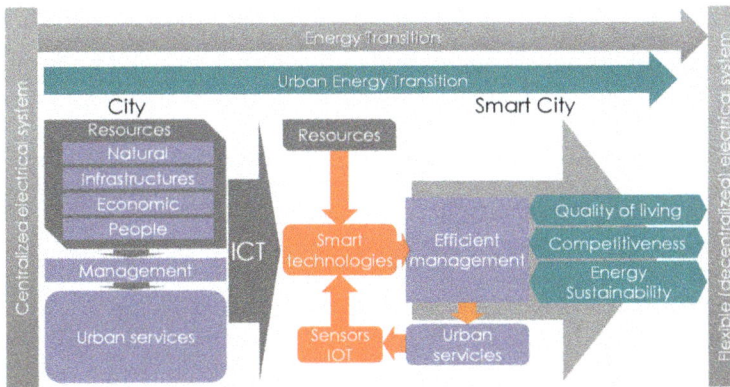

**Figure 4.** Integration of ICT (Information and Communication Technologies) in urban management towards smart cities in the framework of the Urban Energy Transition.

The energy dimension of smart cities is called the smart energy system. Technically, this is a system in which the technological elements of the energy system and the economic agents of the market (consumers, producers, system and market operators, etc.) are combined to identify synergies between them in order to achieve an optimal solution for each individual sector as well as for the overall energy system [10]. Through these synergies, the smart energy systems can reduce the primary energy consumption of cities and cover the remaining demand by taking advantage of the local renewable energy resources, depending on the climatic characteristics of each city and the profitability of the energy vectors [48].

As seen above, the increase in efficiency (however, work on finding a way to increase efficiency is not new for the electricity sector, and, as a result of an industrial process, the electric sector has always been looking for increased efficiency in order to avoid economic losses) is an objective of the flexibilization of the electricity supply value chain through empowering consumers in power generation and management of demand [49]. The main technological components of a flexible electrical system are smart meters, storage systems, distributed generation (DG) systems and smart grids in order to link all of these components together.

Smart grids are electrical distribution grids that, by integrating ICT into their operation, allow for the bidirectional flow between the other technological components of a smart energy system, and the intercommunication between the economic agents of the electric markets, such as consumers, producers, and the system operator [50,51].

Energy transition to an electricity supply without $CO_2$ emissions depends on the deployment of smart grids [52]. These networks are essential for achieving energy security, affordable energy, and climate change mitigation—the three elements of the "energy trilemma" of energy sustainability [53]. The deployment of smart grids has had different objectives in the main economic regions (for the countries of the European Union (and the regulatory proposal "clean energy for all Europeans"), the United States, and Canada, the main drivers in the development of smart grids have been decarbonization and energy efficiency. For Japan and Korea, it has been the "green economy growth agenda". For emerging countries, it has been the rapid growth of their infrastructure [53]) of the world [54]. However, the technical benefits of smart grids will require specific regulation for new electricity market conditions with the entry of new players and incentives for investment in transmission grids [55].

Compared to traditional electricity grids, smart grids allow for the limits of penetration of renewable energies to be exceeded, and grant greater efficiency based on operational control and reliability [56,57]. smart grids ensure an economically efficient and sustainable power system with low losses but high levels of quality, security, of supply and safety [58,59]. The points that are interconnected via the smart grids have double or triple functionality: buildings, lighting systems, and vehicles go from simply consuming energy to being able to generate, store, and export it, thus providing support to the intermittent nature of renewable energies [60]. In this way, smart grids adapt the holistic concept of a smart city in terms of flexibility through smart meters and electrical generation in buildings.

Smart meters are electricity meters that replace traditional electromechanical meters. In front of these, the smart meters measure for the flows of electrical energy imported and/or exported to the grid at time intervals of less than an hour [61]. They also allow the consumer to view this information in real-time via telemetry and obtain new services such as demand side management [62]. This mechanism allows the consumer to respond (demand side response, DSR) to the energy information feedback and non-linear pricing schemes that the market offers. The objective is for the consumer to modify their energy consumption and receive economic benefits in return, while the system benefits in the management of electricity generation. Three of the responses that the consumer can give to these signals are the peak clipping, which is the reduction of the consumption in the peak periods; load shifting, which is the change of the consumption of the peak periods to the off-peak periods; and strategic conservation, which is the change in consumption patterns that reduce energy consumption [63].

### 3. Smart city Strategy to Energy Sustainability

This section reviews the shared aspects of energy sustainability and the smart city urban management strategy arising from the analyses above.

The energy transition in cities seeks to change the current model of centralized generation, which is dependent on the consumption of external energy resources, to a distributed one. In this new distributed system, the smart energy systems make the electricity supply more flexible, increasing the efficiency of the distribution value chain, allowing for the integration of renewable energies into the grid and empowering the consumer in the management of demand. Figure 5 shows the process of decentralization of the electrical system within the smart city urban management model.

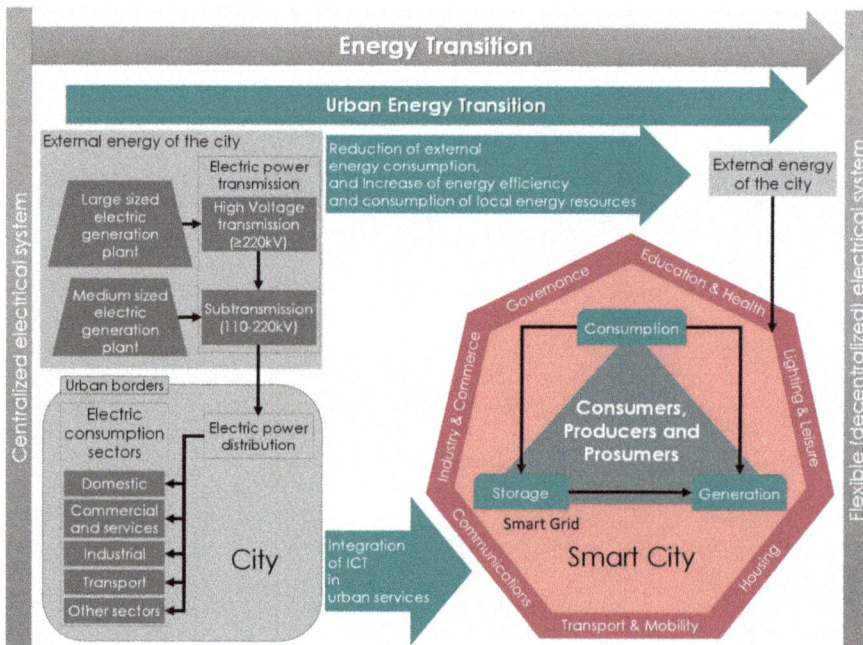

**Figure 5.** Decentralization of the power electric system within the smart city urban management model.

DG can be defined as electric power generation within distribution grids on the customer side of the network [64]. Through this type of generation, it is possible to take advantage of the renewable energy resources in cities. This improves the system efficiency by reducing the losses incurred by the electric transport grid. On-site generation is based on energy self-consumption under the "zero energy" concept, which is a determining factor in the management of energy resources in smart cities.

When applying this concept to the level of buildings, the idea of "zero energy buildings" emerges. These buildings could provide significant and sustainable reductions and help to realize more energy efficient buildings and cities [65]. As shown in Figure 6, the Zero Energy concept is made up of three parts, which are divided based on the balance between demand and the "credits" obtained from energy self-consumption. Nearly zero energy is when demand is low but even so it exceeds self-consumption. Net zero energy occurs when demand equals self-consumption. Plus zero energy is the result of local generation exceeding the demand and the energy can be stored or exported to the grid, depending on market conditions or distribution grid techniques. The advancement of Zero Energy Buildings and ICT in buildings creates smart buildings: Residential buildings, offices, commercial, or industrial buildings and smart homes (homes with a high level of comfort which are technologically and architecturally integrated into their surroundings—also called Green buildings [66]) [67].

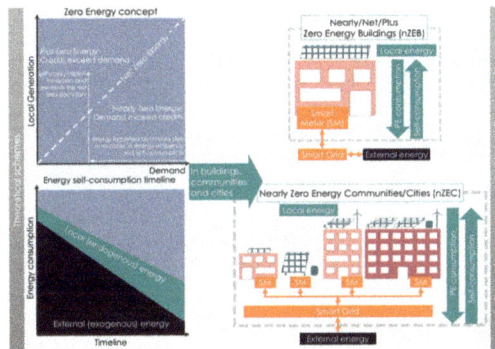

**Figure 6.** Zero energy concept in buildings and cities.

With the presence of smart meters connecting each point of consumption and generation to the smart grid, "zero energy buildings" (Nearly (nZEB), Net (NZEB), and Plus (PZEB)) can share electricity to other DG systems in communities or cities. Therefore, referring back to Figure 6 and taking the "zero energy concept" to a community (building or city) level, it is possible to consider the idea of Nearly Zero Energy Communities/Cities (nZEC); that is to say urban systems made up of "zero energy buildings" and other systems of generation in which, based on the consumption of locally-generated (endogenous) energy, the consumption of external (exogenous) energy is reduced.

From the smart technological development in the distribution grids, new services emerge which will help to increase the energy efficiency of the system by means of the economic incentive to new market agents. Regarding the consumer, although the DSR is a strategy dating from before the development of smart technologies, with smart meters the consumer can manage electric demand in real-time. Similarly, from Zero Energy Buildings, the consumer can become a prosumer of electric energy by adding their surplus energy to the grid and receiving an economic benefit in return. On the other hand, through the advance of distributed generation, PV, and small wind energy have a large margin for growth. This means that energy producers can take advantage of urban spaces or rooftops to generate electricity near consumers. Likewise, demand aggregator comes into play. This is an agent that can bring together the interests of consumers, prosumers and small producers of electrical power to offer services to the system operator [68].

The smart city urban management strategy aims to contribute to the energy sustainability of the planet [6]. However, achieving this objective depends on the progress of DG and; therefore, depends on the investments made in nZEB/NZEB/PZEB towards nZEC [69]. In this sense, the evaluation of investment projects of this type of buildings, and cities, covers the analysis of a wide variety of technical, financial, and economic conditions of the markets [70]. Thus, the advance of generation systems that take advantage of renewable energy resources in cities will produce a whole range of changes for the electricity system, among which are the loss of value of existing infrastructure and the operation of the energy sector [71].

The exploration of the smart city concept indicates that this model or city strategy is presented as the urban management solution that has effectiveness as the aim. Combining "efficacy" and "efficiency", the operation of the smart city energy system seeks:

- Guarantee the efficacy of the energy service in cities, based on the safety and quality of the electricity supply (meaning no interruptions in service) in order to provide services (mobility, lighting, heating, cooling, health, etc.) as demanded by the urban population.
- Manage efficiently the resources needed to provide the energy service to the city: energy resources, economic resources, and the existing infrastructures.

Energy technologies have always focused on efficacy. However, with the superposition of the SMART and ICT criteria, the electric power system gains efficiency to obtain economic and environmental benefits. Figure 7 presents a summary of the results of the analysis of the relationship between effective provision of the energy service and the achievement of the objectives cities and countries have in terms of energy sustainability.

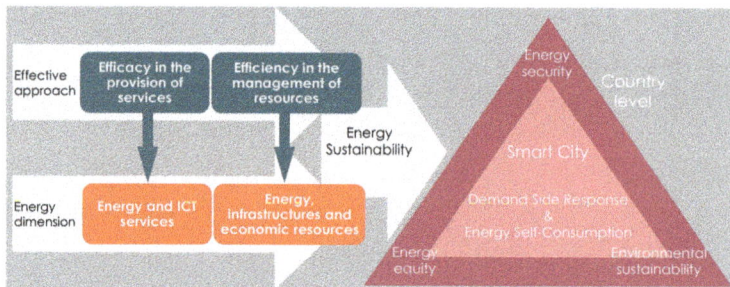

**Figure 7.** Effectiveness of the energy dimension of smart cities to achieve energy sustainability in urban areas and countries.

## 4. Urban Population and Urban PV Generation vs. Energy Sustainability

Because of the urban population increase, cities and countries are closely related. Thus, the relationship between an urban resource management model that aims to contribute to the countries' energy sustainability has been previously described. In this sense, in this part of the paper the relationship between urban population and energy sustainability is analyzed based on empirical data. Based on this, the capacity of local electricity generation that cities have to cover the growing electricity demand, and the effect that this could have on the energy sustainability of the countries, is subsequently analyzed.

### 4.1. Energy Trilemma Index and Urban Population in the Period 2014–2017

Using the results of the ETI during the period 2014–2017, and data from the United Nations [3] on the percentage of urban population, Figure 8 was constructed to analyze the relationship between these two variables. The positions one to 125 of the ETI ranking, organized into five segments of 25 countries, run along the horizontal axis. The vertical axis on the left shows the percentage of population concentration (UP) and the results of ETI standardized between 0–100, while the vertical axis on the right shows the urban population of the five segments. The results obtained are described below.

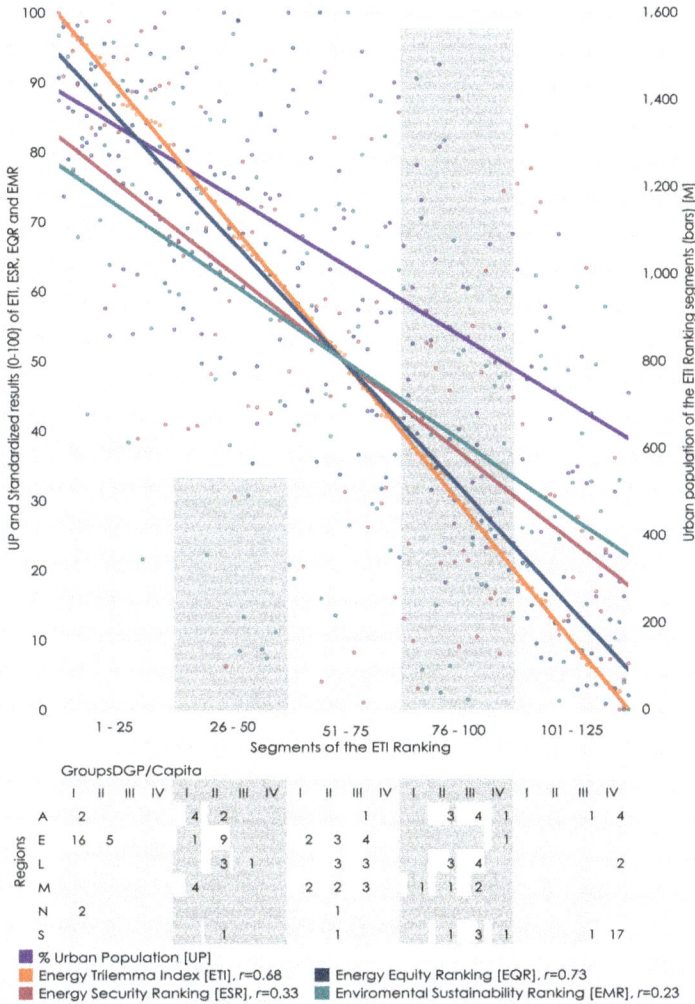

**Figure 8.** Correlation between the percentage of urban population and the ETI during the period 2014–2017. Source: Created by authors using data from [3,12,71–74].

There is a positive correlation between the best results in energy sustainability and the UP of the 125 countries classified in the ETI 2017, as well as with each of its three pillars: From greatest to least, this correlation (*r*) between the UP and EQR, ETI, ESR, and EMR is 0.73, 0.68, 0.33, and 0.23, respectively. The first three of the five country segments (positions 1–75) on average have an UP of above 73% (segments one to three, respectively: 77.4%, 75.86%, and 73.16%). Despite the fourth segment (positions 76–100) having an average UP of 54.17%, it has the highest urban population value (1.56 billion) of all segments and it is almost the sum of the first three segments. As for the fifth segment (positions 101–125), it has the lowest average UP (38.53%) and the lowest value of urban population.

The first 25 countries in the ETI ranking are rich countries in Europe with the highest %UP, over 75%, and the final segment is mostly made up of countries of the Group-IV and have an %UP of below 25%. The second segment is made up of countries from the groups Groups-GDP II and III of Europe, Asia, and MENA. The third segment has a more balanced distribution between countries in Europe, LAC, and MENA of the Groups-GDP I, II, and III. In the fourth segment, the largest in urban population, it is made up of Asian, LAC, and countries of the sub-Saharan region. Most of the latter are found in the fifth segment.

The first 25 countries in the ETI ranking are rich countries in Europe with the highest %UP, over 75%, and the final segment is mostly made up of countries of the Group-IV and have an %UP of below 25%. The second segment is made up of countries from the groups Groups-GDP II and III of Europe, Asia, and MENA. The third segment has a more balanced distribution between countries in Europe, LAC, and MENA of the Groups-GDP I, II, and III. In the fourth segment, the largest in urban population, it is made up of Asian, LAC, and countries of the sub-Saharan region. Most of the latter are found in the fifth segment.

To analyze these results in depth, Figure 9 was constructed, which shows the variation of the 125 countries in their ranking positions and the percentage of urban population (%UP) from 2014 to 2017. This variation is plotted in the four quadrants of a Cartesian axis, in which the horizontal axis represents the UP (growth in the positive axis and decrease in the negative) and the vertical axis represents the variation of the position in the ETI ranking (the higher positions on the positive axis and a lower position on the negative). Each of these four axes is divided into segments of positive correlations ($0.5 > r \leq 1$ and $0 \geq r \leq 0.5$) and negative correlations ($-0.5 \geq r < 0$ and $-1 \geq r > -0.5$). The table on the right shows the percentage of countries in each quadrant in relation to the total number of countries, distributed in each region and Group-GDP.

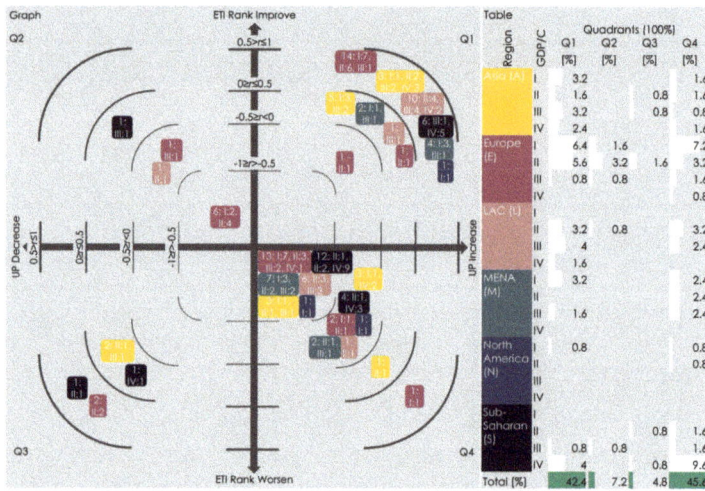

| Region | GDP/C | Q1 [%] | Q2 [%] | Q3 [%] | Q4 [%] |
|---|---|---|---|---|---|
| Asia (A) | I | 3.2 | | | 1.6 |
| | II | 1.6 | | 0.8 | 1.6 |
| | III | 3.2 | | 0.8 | 0.8 |
| | IV | 2.4 | | | 1.6 |
| Europe (E) | I | 6.4 | 1.6 | | 7.2 |
| | II | 5.6 | 3.2 | 1.6 | 3.2 |
| | III | 0.8 | 0.8 | | 1.6 |
| | IV | | | | 0.8 |
| LAC (L) | I | | | | |
| | II | 3.2 | 0.8 | | 3.2 |
| | III | 4 | | | 2.4 |
| | IV | 1.6 | | | |
| MENA (M) | I | 3.2 | | | 2.4 |
| | II | | | | 2.4 |
| | III | 1.6 | | | 2.4 |
| | IV | | | | |
| North America (N) | I | 0.8 | | | 0.8 |
| | II | | | | 0.8 |
| | III | | | | |
| | IV | | | | |
| Sub-Saharan (S) | I | | | | |
| | II | | | 0.8 | 1.6 |
| | III | 0.8 | 0.8 | | 1.6 |
| | IV | | | 0.8 | 9.6 |
| Total (%) | | 42.4 | 7.2 | 4.8 | 45.6 |

**Figure 9.** Variation of the 125 countries in the ETI ranking and the percentage of urban population (%UP) from 2014 to 2017. Source: Created by authors using data from [3,12,71–74].

In 110 of these countries, the urban population percentage increases and, of these, 48% move higher up in the ETI and the remaining 52% go down. There is a large number of countries (57) in the fourth quadrant (Q4) that go down in the ETI rankings but increase their UP; 42 of these countries have a negative correlation between −1 and −0.5 (mainly countries from the regions of Europe and Sub-Saharan), which places them close to the first quadrant (Q1). This quadrant is in second position by number of countries (53), which have gone up in the ETI rankings as well as UP; 43 of these countries have a positive correlation between 0.5 and 1. The European countries of Group-GDP I and II are primarily found in this quadrant. In the second quadrant (Q2), there are nine countries that move up the ETI but their UP decreases. In the third quadrant (Q3), there are six countries that go down in the ETI and have a reduction in UP.

According to these results, in specific terms there is no clear indication that the increase in urban population percentage has a causative effect on the ETI ranking and therefore better energy sustainability. However, 43 countries which do move up in the ranking do so with a high positive correlation (Q1). Furthermore, most of the countries that move down in the ranking do so with a high negative correlation (Q4). This suggests that these countries are close to being in Q1. This indicates that there is a link between a higher ranking in the ETI and the increase of UP.

In conclusion, the concentration of the urban population of a country influences its energy sustainability. The countries with the highest concentration of urban population have a better position in the ETI ranking, and the variation to a better position goes hand in hand with the increase in the percentage of urban population. This implies that the countries respond to their energy sustainability from the management of the energy requirements of their cities.

### 4.2. Photovoltaic Generation in Cities and ETI 2017

As mentioned above, cities are at the center of the energy transition due to their high electricity consumption and the high energy saving capacity of buildings. Additionally, given the uniformity of the distribution of the solar resource on a global scale [75], cities have the capacity to generate electricity near the same consumption points. By deploying PV systems in urbanized areas, cities can cover part of their electricity demand and contribute to the use of local energy resources in their countries. Thus, PV generation is the main technology to move towards the energy sustainability of cities.

PV is the fastest growing renewable energy technology in the last decade. Its generation capacity has grown from 2026 MW in 2009 to 480,357 MW in 2018 [76]. [77] estimates that by 2023 this growth path will double, with a significant acceleration in the growth of distributed generation. Beyond the environmental benefits and energy independence of many regions, the deployment of PV technology could generate EUR 6.67 billion gross value added to Europe and 9 million jobs to 2050 worldwide [78].

The use of solar energy in cities is an objective already underway ([79] reports the commitment of 100 cities in the United States to cover 100% of their energy consumed with renewable energies. [80] reports the same objective for cities of the United Kingdom. [81,82] study the feasibility of this possibility for the entire planet). On a voluntary basis, 65 cities from 56 countries in the world have reported, in 2018 to the CDP (CDP is a not-for-profit charity that runs a disclosure system on environmental data [83]) organization, their objectives for the use of renewable energies to be met in the next 30 years. Thirty-one of these cities include the use of solar energy, and 21 aim to reach 100%.

Based on the great importance of PV generation in the energy transition, this section presents an analysis of the urban PV generation contribution to the energy sustainability of the ETI 2017 countries. The specific objective was to study the effect of the increase of this generation in the indicators "concentration (reduced diversity) of electricity generation", "concentration (reduced diversity) of total PE supply" (PE, primary energy), "import dependence", and "GHG emissions from energy sector" of the pathway calculator (WEC).

The hypothesis followed for this analysis is that the use of the urban PV would allow these countries to obtain a better energy sustainability. Before addressing the description of the calculation and analysis procedure followed, the results obtained in the manipulation of the four pathway calculator indicators from which this hypothesis was raised are described below. Table 1 shows these results.

Reducing the concentration of electricity generation would allow 113 of the 125 countries to obtain a better position in the ETI and; therefore, better energy sustainability. The reduction of the concentration in the total supply of primary energy would allow 106 countries to obtain a better position in the ranking. The reduction of the dependence on energy imports would also allow the 125 countries to obtain a better position in the ranking (only Denmark would retain the same position as in the other three indicators, because this country already has the number 1 position in the ETI ranking). For the last, the reduction of GHG emissions in the energy sector would allow all these countries to obtain a better position in the ranking.

**Table 1.** Results obtained in indicators of the pathway calculator (WEC). Source: Created by authors using data from [8].

| Indicator | ETI Ranking Segment | Variation to the Minimum | | | | Variation to the Maximum | | | |
|---|---|---|---|---|---|---|---|---|---|
| | | Worst | Same | Best | N/A | Worst | Same | Best | N/A |
| Concentration (reduced diversity) of electricity generation (0–100) * | 1–25 | 0 | 9 | 16 | 0 | 25 | 0 | 0 | 0 |
| | 26–50 | 0 | 0 | 25 | 0 | 20 | 5 | 0 | 0 |
| | 51–75 | 1 * | 0 | 24 | 0 | 17 | 8 | 0 | 0 |
| | 76–100 | 0 | 1 | 24 | 0 | 19 | 6 | 0 | 0 |
| | 101–125 | 0 | 1 | 24 | 0 | 16 | 9 | 0 | 0 |
| | Total | 1 | 11 | 113 | 0 | 97 | 28 | 0 | 0 |
| Concentration (reduced diversity) of total PE supply (0–100) ** | 1–25 | 0 | 6 | 19 | 0 | 25 | 0 | 0 | 0 |
| | 26–50 | 0 | 3 | 22 | 0 | 25 | 0 | 0 | 0 |
| | 51–75 | 1 * | 1 | 23 | 0 | 23 | 1 | 1 | 0 |
| | 76–100 | 0 | 1 | 23 | 1 | 21 | 3 | 0 | 1 |
| | 101–125 | 0 | 2 | 19 | 4 | 18 | 3 | 0 | 4 |
| | Total | 1 | 13 | 106 | 5 | 112 | 7 | 1 | 5 |
| Import dependence (1–100) | 1–25 | 0 | 1 | 24 | 0 | 23 | 1 | 1 | 0 |
| | 26–50 | 0 | 0 | 25 | 0 | 22 | 3 | 0 | 0 |
| | 51–75 | 0 | 0 | 25 | 0 | 21 | 2 | 2 | 0 |
| | 76–100 | 0 | 0 | 25 | 0 | 22 | 2 | 1 | 0 |
| | 101–125 | 0 | 0 | 25 | 0 | 21 | 4 | 0 | 0 |
| | Total | 0 | 1 | 124 | 0 | 109 | 12 | 4 | 0 |
| GHG emissions from energy sector (MtCO$_2$e) (0–10,000) *** | 1–25 | 0 | 7 | 18 | 0 | 25 | 0 | 0 | 0 |
| | 26–50 | 0 | 2 | 22 | 1 | 23 | 1 | 0 | 1 |
| | 51–75 | 0 | 0 | 25 | 0 | 24 | 0 | 1 | 0 |
| | 76–100 | 2 ** | 0 | 22 | 1 | 23 | 0 | 1 | 1 |
| | 101–125 | 0 | 1 | 20 | 4 | 17 | 4 | 0 | 4 |
| | Total | 2 | 10 | 107 | 6 | 112 | 5 | 2 | 6 |

* Results of the variation to the minimum: Eleven countries would maintain the same position (Denmark, Sweden, Switzerland, Germany, Finland, Spain, Portugal, Belgium, Romania, Guatemala, Honduras), and only Peru would fall in the ranking. ** Results of the variation to the minimum: Thirteen countries (Benin, Denmark, Dominican Republic, Greece, Kuwait, Lithuania, Netherlands, Niger, Romania, Slovakia, Sweden, Switzerland, and Venezuela) would maintain the same position, and only Peru would fall in the ranking. *** Results of the variation to the minimum: Two countries (Dominican Republic and Iraq) would obtain a lower position, and 10 countries would maintain the same position (Denmark, Greece, Hungary, Netherlands, Niger, Portugal, Romania, Sweden, Switzerland, and United Kingdom).

The empirical procedure followed to calculate the contribution of the urban PV generation to the sustainability of countries was the following. First we sought to calculate the PV generation capacity of the rooftops of the cities (one or two) most representative of the ETI 2017 countries; subsequently, the amount of energy that this solar generation replaced fossil fuel, nuclear, and hydroelectric generation was calculated, in that order, and increasing the country's renewable generation without increasing the balance of its electric mix. Subsequently, between the generation values obtained from each source within the electric mix, the Herfindahl and Hirschman index (HHI) (the HHI is defined as the sum of the squares of the market shares of the firms/sectors/resources, wherein the market shares are expressed as fractions. The result is proportional to the average market share and weighted by market share. Increases in the HHI generally indicate a decrease in competition and an increase of market power, whereas decreases indicate the opposite) was calculated in order to obtain the concentration of the generation. Finally, the results in this index were normalized between 0 and 100 to compare them with the values published for each country in the 2017.

The cities (variable "city" in Equations (1)–(3)) of each country were selected based on their population and economic representation, using world development indicators of the World Bank Group [4]. Data on the PV generation capacity of the rooftops of the cities of the world are not available, so to supplement this information in the calculations, the usable area of the rooftops of the city of Barcelona (BCN) was used as a reference. To be able to analyze the useable rooftop area of each of the selected cities, the ratio 0.0512 was used (see Equation (3)). This is the ratio between the usable rooftop area of Barcelona, a figure obtained from the map of Barcelona's renewable resources map [84,85], and the area of the city. The solar generation capacity data for each of the selected cities were obtained from the global tilted irradiation (GTI) from the Global Solar Atlas of the World Bank Group [86], using a figure of 16% efficiency [87] to calculate the PV generation (see Equation (2)). The equations calculations are the following:

$$PV\_Urban\_Generation_{(Country)} = \sum_{City} PV\_Generation_{City}, \tag{1}$$

$$PV\_Generation_{(City)} = \mu \times GTI_{City} \times Rooftop\_Area_{City}, \tag{2}$$

$$Rooftop\_Area_{(City)} = Area_{City} \times Rooftop\_Area_{BCN} / Area_{BCN}. \tag{3}$$

As a result of this selection of countries and cities, 183 cities were identified from 123 countries (Finland and Iceland were not included because there were no results for them in the GTI indicator): Sixty-seven of these countries were represented by means of one city and 59 with two cities. Data from the International Energy Statistics of the U.S. Energy Information Administration (EIA) [88] were used to calculate the distribution of the electricity generation mix of the study countries. In addition, in the calculation of $CO_2$ emissions, emission factors calculated by [89] (based on data from the International Energy Agency, IEA) were used.

Figure 10 presents the distribution of the electricity generation mix and the hypothetical urban PV generation of each segment of the ETI ranking countries. According to these results, in each segment, generation with fossil fuels has a major presence and the first segment has the lowest hypothetical urban PV generation.

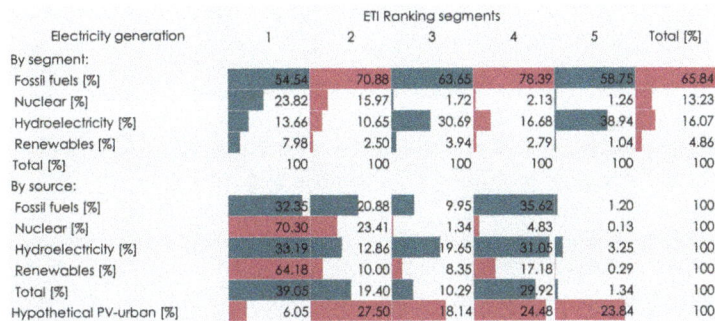

| Electricity generation | ETI Ranking segments | | | | | Total [%] |
|---|---|---|---|---|---|---|
| | 1 | 2 | 3 | 4 | 5 | |
| By segment: | | | | | | |
| Fossil fuels [%] | 54.54 | 70.88 | 63.65 | 78.39 | 58.75 | 65.84 |
| Nuclear [%] | 23.82 | 15.97 | 1.72 | 2.13 | 1.26 | 13.23 |
| Hydroelectricity [%] | 13.66 | 10.65 | 30.69 | 16.68 | 38.94 | 16.07 |
| Renewables [%] | 7.98 | 2.50 | 3.94 | 2.79 | 1.04 | 4.86 |
| Total [%] | 100 | 100 | 100 | 100 | 100 | 100 |
| By source: | | | | | | |
| Fossil fuels [%] | 32.35 | 20.88 | 9.95 | 35.62 | 1.20 | 100 |
| Nuclear [%] | 70.30 | 23.41 | 1.34 | 4.83 | 0.13 | 100 |
| Hydroelectricity [%] | 33.19 | 12.86 | 19.65 | 31.05 | 3.25 | 100 |
| Renewables [%] | 64.18 | 10.00 | 8.35 | 17.18 | 0.29 | 100 |
| Total [%] | 39.05 | 19.40 | 10.29 | 29.92 | 1.34 | 100 |
| Hypothetical PV-urban [%] | 6.05 | 27.50 | 18.14 | 24.48 | 23.84 | 100 |

**Figure 10.** Share of the electricity generation mix and share with urban PV (Photovoltaic) generation by ETI ranking segments 2017. Note: Renewables—biomass and waste, geothermal, solar, tide, and wave electricity, and wind.

The 125 ETI 2017 countries would vary, on average, in the concentration (reduced diversity) of electricity generation by −15.94% due to the hypothetical increase in their urban PV generation. This would have an impact on the reduction of 56.31% of the power electric generation from fossil fuels, and consequently on the reduction of 64% of the $CO_2$ emissions of these countries. According to these results and based on the hypothesis, the urban PV generation of these countries would allow them to improve their energy sustainability (excluding from this conclusion the countries that worsen their position by the variation to the minimum in the four indicators used of the pathway calculator).

Additionally, the use of the local solar resource would allow these countries to improve their sustainability due to the reduction of the dependence on energy imports. Although the diversity (towards concentration) of the primary energy supply would be reduced in some countries, the increase in consumption of renewable resources would allow them to reduce $CO_2$ emissions.

Table 2 summarizes the averages of the results obtained, and Figure 11 presents graphically the results of each segment of countries in the ranking. As indicated in this table, the first segment of the ranking shows the lowest variation of fossil-based electricity generation (−30.93%) and the fifth segment the largest variation (−91.41%): The first segment is occupied by countries with 32.35% of the fossil generation, on the other hand the fifth segment only represents 1.2% of this generation; this indicates that in this segment PV generation has a broader scope to replace fossil-based electricity.

**Table 2.** Average results of the effect of the diversification of the electric generation with urban PV generation in the ETI 2017 countries.

| | | ETI Ranking Segments | | | | |
|---|---|---|---|---|---|---|
| | Total | 1 | 2 | 3 | 4 | 5 |
| Number of countries analyzed | 125 | 25 | 25 | 25 | 25 | 25 |
| Average concentration (reduced diversity) of electricity generation (0–100) | 66.41 | 44.92 | 65.40 | 71.76 | 76.76 | 73.20 |
| After the urban PV generation (0–100) | 55.16 | 28.23 | 44.98 | 55.53 | 61.37 | 85.71 |
| Average variation | −15.94 | −42.49 | −32.33 | −20.91 | −14.29 | 30.32 |
| Number of countries that have diversity electricity generation | 87 | 25 | 21 | 18 | 16 | 7 |
| Average variation of concentration (reduced diversity) [%] | −42.44 | −42.49 | −43.30 | −45.08 | −42.51 | −32.70 |
| Number of countries that concentrate electricity generation | 38 | 0 | 4 | 7 | 9 | 18 |
| Average variation of concentration (reduced diversity) [%] | 44.73 | 0.00 | 25.25 | 41.25 | 35.90 | 54.83 |
| Average variation of the consumption of fossil fuels [%] | −56.31 | −30.93 | −50.24 | −51.08 | −57.88 | −91.41 |
| Average variation in $CO_2$ emission [%] | −64.00 | −40.72 | −59.62 | −61.09 | −65.29 | −93.29 |

The results of the variation in the concentration of electricity generation divide the countries into two groups: A first group of countries that increases the diversity of their electricity generation and a second group of countries that concentrate it.

Countries that diversify their electricity generation (green columns with negative results in Figure 11): A total of 87 countries that, on average, vary the concentration of electricity generation by −42.44%. This means that the initial concentration value of electricity generation decreases due to the increase in PV generation. Within these countries are the 25 countries with the best energy sustainability in the ETI ranking, and with a variation in the concentration of electricity generation (−42.49%) above the total average. On the contrary, the last segment of the ranking is made up of only seven countries and with the lowest value in the variation of the concentration of electricity generation (−32.7%).

Countries which reduce the diversity of their electricity generation (green columns with positive results in Figure 11): This group is made up of 38 countries that on average focus their electricity generation on one single source by 97.57%. This group is comprised of countries in the last segment of the ranking. The diversity of its generation is concentrated in generation from renewable resources, mostly from hydroelectric energy. Although in terms of the energy security ranking, this low level of diversity leads to a lower position in the ETI, the concentration of generation from renewable resources would lead to a reduction in the emissions of GHG and; therefore, a higher position in the environmental sustainability ranking.

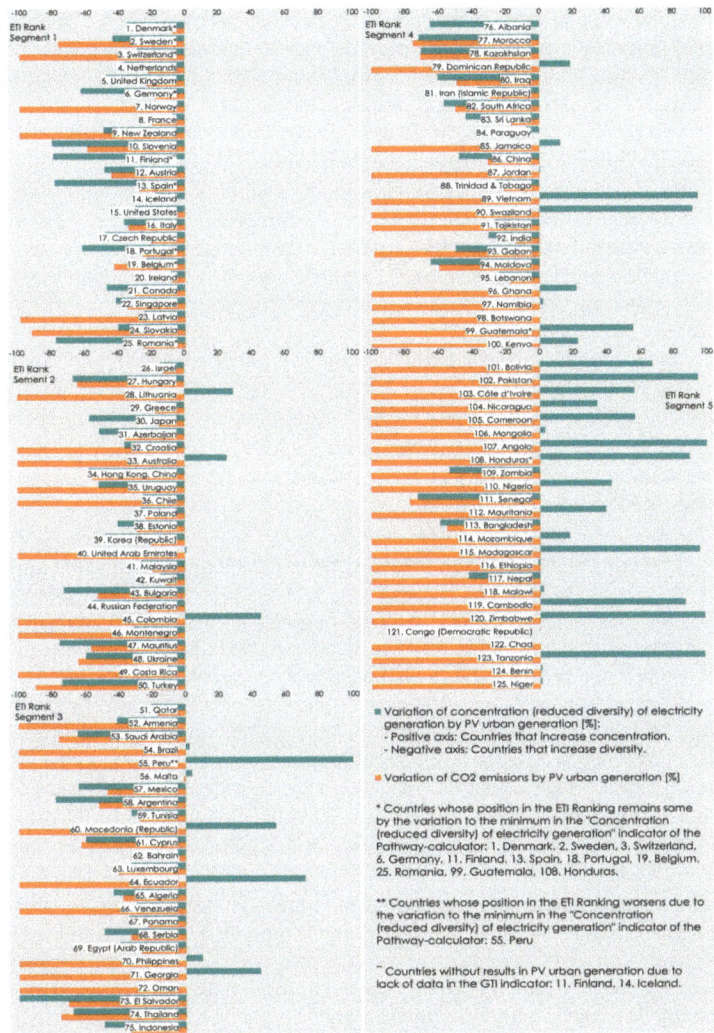

**Figure 11.** Variation in the concentration of electricity generation due to the increase in urban PV generation in the ETI 2017 countries.

## 5. Conclusions

The central axis of this research was the cities: The swarm of urbanization where the population tends to represent the population of the countries lives. Therefore, the energy consumption of cities tends to represent the energy consumption of countries and; therefore, their energy sustainability. On this hypothesis, this paper had three specific objectives: (*i*) To analyze the relationship between the smart urban management strategy and energy sustainability; (*ii*) to analyze the relationship between the urban population increase and the energy sustainability of countries (using the Energy Trilemma Index (ETI) 2014–2017 of the World Energy Council); and (*iii*) to analyze the theoretical self-consumption capacity of PV electricity that cities have to cover the energy demand of the growing population, within the framework of its effect on energy sustainability (using the indicators of the ETI 2017).

The methodology of this research was based on an extensive review of the published literature on the smart city and energy sustainability concepts, and on the use of the results of the ETI for the years 2014–2017 to know a quantitative result of the energy sustainability of countries. Given the results at the country level of this index, and as a contribution to the study of urban energy transition, the research in this paper proposes the analysis of energy sustainability based on the study of the scalability of energy saving and self-consumption from cities.

In conclusion, countries respond to their energy sustainability from the management of the energy requirements of their cities. However, in the face of the increase in the urban population and with the objective of moving towards the decarbonization of their economies, countries can optimize the operation of their urban electric systems through the smart city urban management model. The benefits of applying this strategy also include improving the competitiveness of cities, which leads to economic growth and the development of society.

The main relevant advance of the energy sector in the coming years will be the empowerment of the consumer. Making the electric sector more flexible with the aim of increasing the system's energy efficiency and the penetration of renewable energies relies on the consumer and more specifically the residential sector. One of the major nuclei of energy consumption, and; therefore, party responsible for GHG emissions, is the building. However, it is one of the areas where there is the greatest potential for increasing energy efficiency.

Urban PV generation would have a direct effect on improving the energy sustainability of countries. It would reduce energy dependence, allow for diversifying the mix of electricity generation, and reduce $CO_2$ emissions in the energy sector. Analyzing the three pillars of the ETI has made it possible to identify this effect. Additionally, through this technology, cities have the capacity to respond to the electrical requirements demanded by the growing population. With this, the energy policy of the countries, and mainly those of the countries dependent on external energy resources, should consider the power generation capacity of their cities to avoid increasing the importation of energy. In this sense, although energy security would be affected by the concentration of electric generation in renewable local resources, with a high variability and a difficult prediction, the improvement in environmental issues would be notable. To counteract the imbalance of energy security, smart technologies in cities would allow the management of resources and demand to be optimized.

The inclusion of smart technologies will allow for the integration of the parties involved in the management of the cities, in order to find the balance between the pillars of energy sustainability: energy security, energy equity, and environmental sustainability. Additionally, the smart city model includes a series of elements that will make it possible to provide the energy service citizens require as well as efficiently manage the resources needed to provide them. Specifically, the deployment of smart meters in households and the energy self-consumption of buildings and cities are two key strategies to be able to increase energy saving and efficiency and make the most of the local renewable energy resources.

The energy transition to the decarbonization of the economy through the decentralization of energy systems depends on three main elements of cities: the availability of local renewable energy resources, consumer energy consumption patterns, and regulatory and market characteristics favorable for investment in distributed generation systems. Thus, due to its effect on countries' energy sustainability, the management of current infrastructure and the management of investments in smart distribution systems depends on the national objectives of each economic region in the world.

**Author Contributions:** Conceptualization M.V.-A. and A.S.; methodology, M.V.-A.; investigation, M.V.-A.; writing—original draft preparation, M.V.-A.; writing—review and editing, M.V.-A. and A.S.

**Funding:** This research received no external funding.

**Conflicts of Interest:** The authors declare no conflict of interest.

## References

1. World Commission on Environment and Development. Report of the World Commission on Environment and Development: Our Common Future (The Brundtland Report). *Med. Confl. Surviv.* **1987**, *4*, 300. [CrossRef]
2. IEA (International Energy Agency). *World Energy Outlook-2018*; OECD/IEA: Paris, France, 2018.
3. United Nations: Department of Social and Economic Affairs. *World Population Prospects: The 2012 Revision, Key Findings and Advance Tables*; Working Paper No. ESA/P/WP.227; United Nations: New York, NY, USA, 2013.
4. World Bank. World Development Indicators Data Catalog. Available online: https://datacatalog.worldbank.org/dataset/world-development-indicators (accessed on 21 June 2019).
5. European Climate Foundation. Roadmap 2050: A Practical Guide to a Prosperous, Low-Carbon Europe 2010:3. Available online: https://www.roadmap2050.eu/attachments/files/Roadmap2050-AllData-MinimalSize.pdf (accessed on 21 June 2019).
6. IRENA Renewable Energy in Cities, International Renewable Energy Agency, Abu Dhabi. Available online: www.irena.org/publications (accessed on 21 June 2019).
7. IDC; Anteverti. Hoja de Ruta Para la Smart City. Available online: http://www.ctecno.cat/wp-content/uploads/2012/03/Hoja-de-Ruta-Smart-Cities-def..pdf (accessed on 21 June 2019).
8. Angelidou, M. Smart city policies: A spatial approach. *Cities* **2014**, *41*, S3–S11. [CrossRef]
9. Espinoza-Arias, P.; Poveda-Villalón, M.; García-Castro, R.; Corcho, O. Ontological Representation of Smart City Data: From Devices to Cities. *Appl. Sci.* **2018**, *9*, 32. [CrossRef]
10. Lund, H.; Østergaard, P.A.; Connolly, D.; Mathiesen, B.V. Smart energy and smart energy systems. *Energy* **2017**, *137*, 556–565. [CrossRef]
11. World Energy Council (WEC) and Oliver Wyman. Energy Trilemma Index 2013. Available online: https://www.worldenergy.org/wp-content/uploads/2013/09/2013-Energy-Sustainability-Index-VOL-2.pdf (accessed on 21 June 2019).
12. World Energy Council (WEC) and Oliver Wyman. Energy Trilemma Index 2017. Available online: https://www.worldenergy.org/wp-content/uploads/2017/11/Energy-Trilemma-Index-2017-Report.pdf (accessed on 21 June 2019).
13. World Energy Council (WEC) and Oliver Wyman. Pathway Calculator. 2017. Available online: https://trilemma.worldenergy.org/#!/pathway-calculator (accessed on 28 May 2019).
14. Doran, G.T. There's a S.M.A.R.T. way to write management's goals and objectives. *Manag. Rev.* **1981**, *70*, 35–36.
15. Lee, J.; Lee, H. Developing and validating a citizen-centric typology for smart city services. *Gov. Inf. Q.* **2014**, *31*, S93–S105. [CrossRef]
16. Caragliu, A.; Del Bo, C.; Nijkamp, P. Smart Cities in Europe. *J. Urban Technol.* **2011**, *18*, 65–82. [CrossRef]
17. Oliviera Fernandes, E.; Meeus, L.; Leal, V.; Azevedo, I.; Delarue, E.; Glachant, J.-M. *Smart Cities Initiative: How to Foster a Quick Transition Towards Local Sustainable Energy Systems*; European University Institute: Firenze, Italy, 2011.
18. Batty, M.; Axhausen, K.W.; Giannotti, F.; Pozdnoukhov a Bazzani a Wachowicz, M.; Ouzounis, G.; Portugali, Y. Smart cities of the future. *Eur. Phys. J. Spec. Top* **2012**, *214*, 481–518. [CrossRef]
19. Kramers, A.; Höjer, M.; Lövehagen, N.; Wangel, J. Smart sustainable cities—Exploring {ICT} solutions for reduced energy use in cities. *Environ. Model. Softw.* **2014**, *56*, 52–62. [CrossRef]
20. Caragliu, A.; DBo, C.; Kourtit, K.; Nijkamp, P. *Smart Cities*; Elsevier: Amsterdam, The Netherlands, 2015. [CrossRef]
21. Gonz, L. El Papel De Las Normas en Las Ciudades Inteligentes. 2014. Available online: https://www.esmartcity.es/biblioteca/informe-el-papel-de-las-normas-en-las-ciudades-inteligentes (accessed on 21 June 2019).
22. Manville, C.; Cochrane, G.; Cave, J.; Millard, J.; Pederson, J.K.; Thaarup, R.K.; Liebe, A.; Wissner, M.; Massink, R.; Kotterink, B. *Mapping Smart Cities in the EU*; European Parliament: Brussels, Belgium, 2014; Volume 200. [CrossRef]
23. Achaerandio, R.; Galloti, G.; Curto, J.; Bigliani, R.; Maldonado, F. Análisis de las Ciudades Inteligentes en España. 2011. Available online: https://www.esmartcity.es/biblioteca/analisis-de-idc-de-las-ciudades-inteligentes-en-espana (accessed on 21 June 2019).

24. Drohojowska, H. San Francisco Style, Art-Deco Elements Inform a Smart City Residence + Interior-Design by Arnold, Val. *Archit. Dig.* **1991**, *48*, 114–121.

25. Shetty, V. A tale of smart cities. *Commun. Int.* **1997**, *24*, 16–18.

26. Neirotti, P.; De Marco, A.; Cagliano, A.C.; Mangano, G.; Scorrano, F. Current trends in Smart City initiatives: Some stylised facts. *Cities* **2014**, *38*, 25–36. [CrossRef]

27. Albino, V.; Berardi, U.; Dangelico, R.M. Smart Cities: Definitions, Dimensions, Performance, and Initiatives. *J. Urban Technol.* **2015**, *22*, 3–21. [CrossRef]

28. Angelidou, M. Smart cities: A conjuncture of four forces. *Cities* **2015**, *47*, 95–106. [CrossRef]

29. Studies, M. *Smart Cities Ranking of European Medium-Sized Cities*; Vienna University of Technology: Vienna, Austria, 2007; Volume 16, pp. 13–18.

30. Lazaroiu, G.C.; Roscia, M. Definition methodology for the smart cities model. *Energy* **2012**, *47*, 326–332. [CrossRef]

31. Mattoni, B.; Gugliermetti, F.; Bisegna, F. A multilevel method to assess and design the renovation and integration of Smart Cities. *Sustain. Cities Soc.* **2015**, *15*, 105–119. [CrossRef]

32. Ben Letaifa, S. How to strategize smart cities: Revealing the SMART model. *J. Bus. Res.* **2015**, *68*, 1414–1419. [CrossRef]

33. Lee, J.H.; Hancock, M.G.; Hu, M.-C. Towards an effective framework for building smart cities: Lessons from Seoul and San Francisco. *Technol. Forecast. Soc. Chang.* **2013**, *89*, 80–99. [CrossRef]

34. Bakici, T.; Almirall, E.; Wareham, J. A Smart City Initiative: The Case of Barcelona. *J. Knowl. Econ.* **2013**, *4*, 135–148. [CrossRef]

35. Lee, J.H.; Phaal, R.; Lee, S.-H. An integrated service-device-technology roadmap for smart city development. *Technol. Forecast. Soc. Chang.* **2013**, *80*, 286–306. [CrossRef]

36. European Commission. *A Resource-Efficient Europe—Flagship Initiative Under the Europe 2020 Strategy*; European Commission: Brussels, Belgium, 2011.

37. European Commission. *Smart Cities and Communities—European Innovation Partnership*; European Commission: Brussels, Belgium, 2012.

38. Ametic. Smart Cities 2013. Available online: https://www.hr.com/en/communities/2012-quality-of-living-worldwide-city-rankings-%E2%80%93-m_haderw3e.html (accessed on 21 June 2019).

39. Harrison, C.; Donnelly, I.A. A Theory of Smart Cities. In Proceedings of the 55th Annual Meeting of the International Society for the Systems Sciences, Houcheng, UK, 17–22 July 2011.

40. Sanchez, L.; Muñoz, L.; Galache, J.A.; Sotres, P.; Santana, J.R.; Gutierrez, V.; Ramdhany, R.; Gluhak, A.; Krco, S.; Theodoridis, E.; et al. SmartSantander: IoT experimentation over a smart city testbed. *Comput. Netw.* **2014**, *61*, 217–238. [CrossRef]

41. Mercer. *2012 Quality Of Living Worldwide City Rankings: Survey*; International HR Adviser: Sutton, UK, 2012; pp. 33–36.

42. Paroutis, S.; Bennett, M.; Heracleous, L. A strategic view on smart city technology: The case of IBM Smarter Cities during a recession. *Technol. Forecast. Soc. Chang.* **2013**, *89*, 262–272. [CrossRef]

43. Talari, S.; Shafie-Khah, M.; Siano, P.; Loia, V.; Tommasetti, A.; Catalão, J.P.S. A review of smart cities based on the internet of things concept. *Energies* **2017**, *10*, 421. [CrossRef]

44. Mahapatra, C.; Moharana, A.K.; Leung, V.C.M. Energy management in smart cities based on internet of things: Peak demand reduction and energy savings. *Sensors* **2017**, *17*, 2812. [CrossRef] [PubMed]

45. Hilty, L.M.; Aebischer, B.; Rizzoli, A.E. Modeling and evaluating the sustainability of smart solutions. *Environ. Model. Softw.* **2014**, *56*, 1–5. [CrossRef]

46. European Commission. *Energy Technologies and Innovation*; European Commission: Brussels, Belgium, 2013.

47. Höjer, M.; Wangel, J. Smart Sustainable Cities Definition and Challenges. *ICT Innov. Sustain.* **2014**, 333–349. [CrossRef]

48. Lund, H.; Werner, S.; Wiltshire, R.; Svendsen, S.; Thorsen, J.E.; Hvelplund, F.; Mathiesen, B.V. 4th Generation District Heating (4GDH): Integrating smart thermal grids into future sustainable energy systems. *Energy* **2014**, *68*, 1–11. [CrossRef]

49. IRENA. Power System Flexibility for the Energy Transition, International Renewable Energy Agency, Abu Dhabi. 2018. Available online: www.irena.org/publications (accessed on 21 June 2019).

50. Siano, P. Demand response and smart grids—A survey. *Renew. Sustain. Energy Rev.* **2014**, *30*, 461–478. [CrossRef]

51. Yuan, J.; Shen, J.; Pan, L.; Zhao, C.; Kang, J. Smart grids in China. *Renew. Sustain. Energy Rev.* **2014**, *37*, 896–906. [CrossRef]

52. Solomon, B.D.; Krishna, K. The coming sustainable energy transition: History, strategies, and outlook. *Energy Policy* **2011**, *39*, 7422–7431. [CrossRef]

53. World Energy Council (WEC). *Smart Grids: Best Practice Fundamentals for a Modern Energy System*; WEC: London, UK, 2012; ISBN 9780946121175. Available online: https://www.worldenergy.org/wp-content/uploads/2012/10/PUB_Smart_grids_best_practice_fundamentals_for_a_modern_energy_system_2012_WEC.pdf (accessed on 21 June 2019).

54. Dada, J.O. Towards understanding the benefits and challenges of Smart/Micro-Grid for electricity supply system in Nigeria. *Renew. Sustain. Energy Rev.* **2014**, *38*, 1003–1014. [CrossRef]

55. Clastres, C. Smart grids: Another step towards competition, energy security and climate change objectives. *Energy Policy* **2011**, *39*, 5399–5408. [CrossRef]

56. Lund, P.D.; Mikkola, J.; Ypyä, J. Smart energy system design for large clean power schemes in urban areas. *J. Clean. Prod.* **2014**, *103*, 437–445. [CrossRef]

57. Ruiz-Romero, S.; Colmenar-Santos, A.; Mur-Pérez, F.; López-Rey, Á. Integration of distributed generation in the power distribution network: The need for smart grid control systems, communication and equipment for a smart city—Use cases. *Renew. Sustain. Energy Rev.* **2014**, *38*, 223–234. [CrossRef]

58. Gaiser, K.; Stroeve, P. The impact of scheduling appliances and rate structure on bill savings for net-zero energy communities: Application to West Village. *Appl. Energy* **2014**, *113*, 1586–1595. [CrossRef]

59. Ellabban, O.; Abu-Rub, H. Smart grid customers' acceptance and engagement: An overview. *Renew. Sustain. Energy Rev.* **2016**, *65*, 1285–1298. [CrossRef]

60. Mwasilu, F.; Justo, J.J.; Kim, E.-K.; Do, T.D.; Jung, J.-W. Electric vehicles and smart grid interaction: A review on vehicle to grid and renewable energy sources integration. *Renew. Sustain. Energy Rev.* **2014**, *34*, 501–516. [CrossRef]

61. Uribe-Pérez, N.; Hernández, L.; de la Vega, D.; Angulo, I. State of the Art and Trends Review of Smart Metering in Electricity Grids. *Appl. Sci.* **2016**, *6*, 68. [CrossRef]

62. Goulden, M.; Bedwell, B.; Rennick-Egglestone, S.; Rodden, T.; Spence, A. Smart grids, smart users? the role of the user in demand side management. *Energy Res. Soc. Sci.* **2014**, *2*, 21–29. [CrossRef]

63. Eid, C.; Koliou, E.; Valles, M.; Reneses, J.; Hakvoort, R. Time-based pricing and electricity demand response: Existing barriers and next steps. *Util. Policy* **2016**, *40*, 15–25. [CrossRef]

64. Ackermann, T.; Andersson, G.; Söder, L. Distributed generation: A definition. *Electr. Power Syst. Res.* **2001**, *57*, 195–204. [CrossRef]

65. Chen, J.; Jain, R.K.; Taylor, J.E. Block Configuration Modeling: A novel simulation model to emulate building occupant peer networks and their impact on building energy consumption. *Appl. Energy* **2013**, *105*, 358–368. [CrossRef]

66. GhaffarianHoseini, A.; Dahlan, N.D.; Berardi, U.; GhaffarianHoseini, A.; Makaremi, N.; GhaffarianHoseini, M. Sustainable energy performances of green buildings: A review of current theories, implementations and challenges. *Renew. Sustain. Energy Rev.* **2013**, *25*, 1–17. [CrossRef]

67. GhaffarianHoseini, A.; Dahlan, N.D.; Berardi, U.; GhaffarianHoseini, A.; Makaremi, N. The essence of future smart houses: From embedding {ICT} to adapting to sustainability principles. *Renew. Sustain. Energy Rev.* **2013**, *24*, 593–607. [CrossRef]

68. IRENA. Innovation Landscape: Aggregators, International Renewable Energy Agency, Abu Dhabi. 2019. Available online: www.irena.org/publications (accessed on 21 June 2019).

69. Villa-Arrieta, M.; Sumper, A. Economic evaluation of Nearly Zero Energy Cities. *Appl. Energy* **2019**, *237*, 404–416. [CrossRef]

70. Villa-Arrieta, M.; Sumper, A. A model for an economic evaluation of energy systems using TRNSYS. *Appl. Energy* **2018**, *215*, 765–777. [CrossRef]

71. OECD/IEA; IRENA. Perspectives for the Energy Transition: Investment Needs for a Low-Carbon Energy System, Abu Dhabi. 2019. Available online: www.irena.org/publications (accessed on 21 June 2019).

72. World Energy Council (WEC) and Oliver Wyman. Energy Trilemma Index 2014. Available online: https://www.worldenergy.org/publications/2014/world-energy-trilemma-2014-time-to-get-real-the-myths-and-realities-of-financing-energy-systems/ (accessed on 21 June 2019).

73. World Energy Council (WEC) and Oliver Wyman. Energy Trilemma Index 2015. Available online: https://www.worldenergy.org/publications/2015/2015-energy-trilemma-index-benchmarking-the-sustainability-of-national-energy-systems-2/ (accessed on 21 June 2019).

74. World Energy Council (WEC) and Oliver Wyman. Energy Trilemma Index 2016. Available online: https://www.worldenergy.org/publications/2016/2016-energy-trilemma-index-benchmarking-the-sustainability-of-national-energy-systems/ (accessed on 21 June 2019).

75. Check, R.; Space-based, P.S.; Percent, K.; Shingles, E. *The Future of Solar Energy*; MIT Energy Initiative: Cambridge, MA, USA, 2015; pp. 3–6.

76. IRENA. Renewable Energy Capacity Statistics 2019, International Renewable Energy Agency, Abu Dhabi. 2019. Available online: www.irena.org/publications (accessed on 21 June 2019).

77. International Energy Agency (IEA). Renewables 2018. *Market Analysis and Forecast from 2018 to 2023*. 2019. Available online: https://www.iea.org/renewables2018/power/ (accessed on 13 June 2019).

78. International Renewable Energy Agency. Renewable Energy Benefits Leveraging Local Capacity for Solar Pv, Abu Dhabi. 2019. Available online: www.irena.org/publications (accessed on 21 June 2019).

79. 100 US Cities Are Committed to 100 Percent Clean, Renewable Energy|Sierra Club 2018. Available online: https://www.sierraclub.org/press-releases/2019/03/100-us-cities-are-committed-100-percent-clean-renewable-energy (accessed on 12 June 2019).

80. UK100 2019. Available online: https://www.uk100.org/ (accessed on 12 June 2019).

81. Ram, M.; Dmitrii, B.; Arman, A.; Solomon, O.; Ashish, G.; Michael, C.; Christian, B. *Global Energy System Based on 100% Renewable Energy Power Sector*; Lappeenranta University of Technology and Energy Watch Group: Lappeenranta, Finland; Berlin, Germany, 2017.

82. Couture, T.D.; Leidreiter, A. *How to Achieve 100% Renewable Energy*; The World Future Council: Hamburg, Germany, 2014.

83. CDP. Home—CDP 2019. Available online: https://www.cdp.net/en (accessed on 19 June 2019).

84. Agencia de Energía de Barcelona. ¿Cuánta Energía Puedes Generar? 2016. Available online: http://energia.barcelona/es/cuanta-energia-puedes-generar (accessed on 20 March 2018).

85. Agencia de Energía de Barcelona. Mapa de Recursos d ' Energia Renovable de Barcelona. 2016. Available online: http://energia.barcelona/ca/quanta-energia-pots-generar (accessed on 20 March 2018).

86. ESMAP; SOLARGIS; WB; IFC. Global Solar Atlas. Glob Sol Atlas 2019:1. Available online: https://globalsolaratlas.info/?c=32.651516,51.678658,11&s=32.65139,51.67917&m=sg:ghi (accessed on 22 May 2019).

87. (ISE) FI for SES. *Photovoltaics Report*; Fraunhofer ISE: Freiburg, Germany, 2016.

88. U.S. Energy Information Administration. International Energy Statistics n.d. Available online: https://www.eia.gov/beta/international/data/browser/#/?c=4100000002000060000000000000g000200000000000000001&vs=INTL.44-1-AFRC-QBTU.A&vo=0&v=H&start=1980&end=2016 (accessed on 21 June 2019).

89. Ecometrica. Electricity-specific emission factors for grid electricity. *Ecometrica* **2011**, 1–22. [CrossRef]

MDPI

St. Alban-Anlage 66

4052 Basel

Switzerland

Tel. +41 61 683 77 34

Fax +41 61 302 89 18

www.mdpi.com

*Applied Sciences* Editorial Office

E-mail: applsci@mdpi.com

www.mdpi.com/journal/applsci

www.ingramcontent.com/pod-product-compliance
Lightning Source LLC
Chambersburg PA
CBHW051916210326
41597CB00033B/6164